Secrets of the Sun

Ronald G. Giovanelli

Commonwealth Scientific and Industrial Research Organization,

Division of Applied Physics

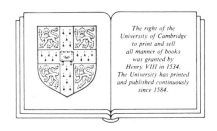

The right of the
University of Cambridge
to print and sell
all manner of books
was granted by
Henry VIII in 1534.
The University has printed
and published continuously
since 1584.

Cambridge University Press

Cambridge

London   New York   New Rochelle

Melbourne   Sydney

Published by the Press Syndicate of the University of Cambridge
The Pitt Building, Trumpington Street, Cambridge CB2 1RP
32 East 57th Street, New York, NY 10022, USA
296 Beaconsfield Parade, Middle Park, Melbourne 3206, Australia

© Cambridge University Press 1984

First published 1984

Printed in Great Britain at the Cambridge University Press

Library of Congress catalogue card number: 84-7687

British Library Cataloguing in Publication Data

Giovanelli, Ronald G.
Secrets of the sun.
1. Sun
I. Title
523.7   QB521

ISBN 0 521 25521 X

Contents

Preface

This is a book about the Sun for non-specialists and non-scientists, for those who do not understand scientific jargon or mathematical equations. It aims to show in pictures and explain in simple language the fascination and paradoxes of the Sun. Some scientific terms are impossible to avoid, but as far as possible these are described in non-scientific language.

Scientists also should appreciate a book for laymen. Your author has endured far too many a learned, dreary lecture by eminent specialists not to realize that he and his long-suffering scientist colleagues also long for simple explanations in simple terms; the simpler the better. It is only in a highly specialized context that the subtleties, the precision of science can be properly appreciated.

Worldwide, science is funded very largely by taxpayers, non-scientists. The non-scientist has as much right as the specialist to enjoy the products of science. When it comes to engineering applications, he gains a tangible reward. But there is another reward which he is too often denied, the aesthetic enjoyment of the subject, the thrill of discovery and the excitement of the search for the explanation. This book cannot set out to provide all these, but it does supply some of them.

Scientists show remarkable cooperation in providing their best photographs and diagrams for a book for non-scientists. The author is extremely grateful to friends and colleagues in many countries who have helped unstintedly. They include J. Beckers, R. J. Bray, Carol Clarke, L. Cram, R. B. Dunn, T. Duvall, E. Frazier, V. Gaizauskas, J. W. Harvey, K. Harvey, R. Howard, A. Krieger, L. Lacey, W. C. Livingston, R. Loughhead, Marie McCabe, B. Pagel, J. Piddington, A. K. Pierce, N. Sheeley, R. Smartt, B. A. Smith, E. Tappere, K. Sheridan, Virginia Sonett, D. Vrabec, J. P. Wild, B. Yallop, H. Zirin and specialist technicians in many institutions, R. Faller, H. Gillett, M. Hanna, R. Rattle, G. Pickens, Myrna Cook, Pauline Gorenc and Deborah Brown.

Several patient critics, Elizabeth Maggio, Dorothy Braxton and Hal Myers have helped the author at various stages in his gropings towards a readable account. The remaining faults, big or small, are mine alone.

Ronald G. Giovanelli
1983

Introduction

It all began with the development of the telescope.

The art of fabricating spectacles by grinding and polishing spherical surfaces on glass developed late in the 1200s, although we can infer that the Minoans knew of magnifying glasses some 2500 years earlier. By the early 1500s, convex and concave spectacle lenses were available widely in Europe. Particularly in the latter half of the century there was much writing, and no doubt talk, of using single lenses or combinations of lenses to see distant objects in great detail. Authors implied that such devices existed, but this was mainly imagination.

The first reliable evidence of the construction of a telescope was contained in a patent application in August 1608 by Hans Lippershey, of Middelburg in Holland. The system used a convex and a concave lens, spaced appropriately. The patent was not allowed, apparently because other Dutchmen, particularly Jacob Metius, claimed to have made telescopes already.

The early Dutch instruments were of low magnification, but the news of their development spread widely and rapidly. The following year Galileo, then at the University in Padua, heard of the telescope. Quickly he worked out sufficient of the basic ideas to construct several of successively greater powers. In 1611, stimulated by Galileo's successes, Johannes Kepler in Prague published the first theory of telescopic combinations of lenses. Although this was sent to Galileo in enthusiasm, he had no apparent need of the wider range of possibilities described and tended to scorn them. They played no part in telescope development until several years later.

Using his telescopes Galileo discovered that the Milky Way was composed of myriads of individual stars, that the Moon's surface was covered in many parts with mountains and craters, that Venus was not a fully illuminated disk but that its bright parts exhibited phases like those of the Moon, that Jupiter had at least four satellites (moons) which orbited the giant planet (convincing him that the planets, too, orbited the Sun, as Copernicus had proposed), and that there were dark spots on the Sun. By early 1611 he was demonstrating these to colleagues in Rome.

While Galileo had produced the best telescopes, he was not the only one to turn the new instruments onto the Sun. In March 1611, Fr Christopher Scheiner in Southern Germany was observing sunspots, but whereas today seeing usually implies believing, the consensus of learned opinion was somewhat different then. It is reported by the nineteenth-century astronomer Richard Proctor that when Scheiner advised his Jesuit superior of these discoveries he was told, 'I have read Aristotle's writings from beginning to end, and I can assure you that I have nowhere found in them anything similar to what you mention; go therefore, my son; tranquillize yourself; be assured that what you take for spots in the Sun are the faults of your glasses or your eyes'. Scheiner was unconvinced. He undertook a detailed investigation of sunspot behaviour which has recently proved to be of great importance for our knowledge of sunspot cycles and solar rotation preceding the Little Ice Age (c. 1650–1700).

Spots had been noted on the Sun before

the telescope, the first known reference being by Theophrastus, of Athens, around 350 BC. Many spots large enough to be seen by the unaided eye were recorded in China from 28 BC onwards (perhaps as early as 165 BC), some for six to 13 consecutive days. None of these records was available in English or any other European language until 1873. Other spots had been recorded in Europe at various times from 807 onwards (840, 1096, 1365, 1371 and 1588) and in Peru in Prehispanic times. Generally, European authorities believed them to be unknown planets close to the Sun, as did Scheiner initially.

In December 1610 or early 1611, Johannes Fabricius in Germany observed sunspots with a camera obscura, or pinhole camera. In this device, light passing in a straight line through a small hole into a darkened room forms an illuminated patch reproducing the source of light: sky, clouds, buildings, etc. The Sun forms a bright circular disk on which dark spots can sometimes be seen (Figs 1 and 13). Fabricius identified them as of solar origin. Because of these identifications, Fabricius and Galileo must be awarded the joint honour of their discovery.

Solar astronomy began with the telescope. But most of the ideas on solar physics during the next 200 years or so would now be called nonsense. This provides a fine example of the supreme mental effort involved in learning about a remote object when the relevant laws of physics are unknown. Sir William Herschel for example, regarded long after his day as the greatest observational astronomer of all time, argued strongly in 1794 that solar radiation originated in a hot thin cloud layer, that occasional gaps in the cloud enabled us to see cooler regions below, and that these appeared as sunspots. Herschel had no doubt that the Sun was solid and protected from the intense heat by a lower level of cloud and he considered that the Sun was probably inhabited! Such was Herschel's authority that many astronomers clung to his views for over 50 years afterwards.

It seems extraordinary, too, that it took 200 years of sunspot observations, at times sporadic, at times systematic, to find the sunspot cycle. It was not until 1843 that Heinrich Schwabe could announce the existence of a cycle of activity about 10 years long (we now know it to be more like 11 years but rather variable) during which sunspot activity rose from nearly zero to a maximum and down to nearly zero again.

Modern solar physics began shortly after Herschel's fantastic assertions. It had long been known that white light was spread out into its component colours, the spectrum, when passing through a prism. The light from today's tungsten filament lamps shows quite smooth changes across the spectrum, but in 1802 William Wollaston, in London, discovered some very narrow colour-ranges missing from sunlight. From their appearance in spectrographs they are called spectral lines (see Chapter 3 and Plates IV and V). Studied in some detail over the next 10 years by Joseph von Fraunhofer of Munich, they were shown to be due to absorption of light by the various chemical elements in the Sun's atmosphere. Before long, this was followed by the identification of the elements present in the Sun; these are the same as are on Earth. Subsequent analysis has shown that the *relative* concentrations are not so very different, with two major exceptions; about 92% of the Sun's atoms are hydrogen, about 8% helium. The other elements make up only about 0.1% of the total – they are almost impurities. To a large extent the Sun is a ball of hydrogen.

In 1869 the American Jonathan Lane initiated the concept that the Sun was a spherical concentration of nothing but gas, held together by gravitation and having a central energy source. His brilliant, revolutionary ideas were further developed around the turn of the century by Robert Emden, a Swiss astrophysicist working in Munich. By 1926 Sir Arthur Eddington had concluded that the energy source must be of nuclear origin. Its specific nature, the fusion of hydrogen atoms to form helium, had to await the 1940s for detailed identification. It was obvious from Emden's theory, and seemed to be so from the principles of thermodynamics, that temperatures on the Sun should decrease outwards continuously; but discoveries by Walter Grotrian in 1939 and Bengt Edlén 3 years later showed that temperatures *increased* outwards in the upper atmosphere to some 1 million K,* destroying such comfortable dogma with inescapable facts, just as had happened in the discovery of sunspots and the Jovian planets over 300 years earlier. Scientists now believe that heating of the Sun's upper atmosphere is due to dissipation of waves propagating outwards from below, but the details are far from clear.

Photographic observations of the surface of the Sun were not made until the 1870s, before which all observations were visual. Sir William Herschel seems to have been the first to become interested in the fine structure of the surface (1801). Little other notice was taken until around 1862 when James Nasmyth, an English amateur, manufactured a sizeable telescope of 20 inches (about 50 cm) aperture. His reports aroused the interests of others, who all saw fine structure but with little agreement as to its properties.

Pierre Janssen's magnificent photographs of solar granulation, obtained at the Meudon Observatory near Paris, should have settled the controversies in the 1880s. Some showed patterns of small, bright structures typically about 2000 kilometres (km) across, surrounded by dark lanes. The majority showed large-scale distortions, now known to be due to variable refraction in the Earth's atmosphere although Janssen believed that they were of solar origin. So it was not until 1957 that photographs obtained by Martin Schwarzschild with a balloon-borne telescope showed incontrovertibly the structure of granulation.

Higher levels in the solar atmosphere can be observed during eclipses, when the disk of the Sun is obscured by the Moon. The extremely faint corona can then be seen, extending far into surrounding space. Eclipses are rare, however, and other techniques have been developed for studying

* Temperatures are quoted throughout on the Kelvin, or absolute, scale and are denoted by K. This equals temperature in Celsius (or centigrade), °C, plus 273.15. Thus 100 °C = 373.15K; 5000 °C = 5273.15K. Fahrenheit temperatures are converted to Kelvin by dividing by 1.8 and adding 255.37. Thus 100 °F = 310.93K. At the high temperatures found in the Sun, Fahrenheit temperatures are roughly 1.8 times the Kelvin temperatures in this book.

the corona and an intermediate region, the chromosphere. Fine structures, often of short life, fill these regions, giving the Sun a weather pattern more complex than ours on Earth, and storms of enormous intensity. The most spectacular is the flare, of which the smallest may last for only a few minutes or less and the biggest for an hour or more. The very biggest flares emit more energy in this time than in all the solar radiation received on the Earth in 300 years!

Today there are five basic problems in solar physics. The first is the sunspot cycle, which is closely interwoven with the second, the structure of the convection zone (occupying about the outer 200 000 km (130 000 miles) of the solar radius), and with the third, the variation of rotation rate across the surface and with depth. The final two are the heating mechanisms in the outer solar layers, and the causes of flares. There are very many other intriguing problems, but central to almost all is the nature of the Sun's magnetic field.

To study all solar problems we must fall back almost inevitably on the spectrum. Following Fraunhofer's work on atomic absorption lines, the relevant theory developed gradually. Major progress came with the development of quantum mechanics in the 1920s, which enabled many of the necessary atomic properties to be calculated and understood. Even so, the measurable properties of the spectral lines depend in an intricate way on many physical properties of the atmosphere, e.g. temperature, density, motions, magnetic fields, and their spatial variations. It has been a task of extreme difficulty to unravel these physical properties from spectral-line observations, but much

progress has been made, as will be shown in this book.

Solar astronomy today is not a cheap science. Many of the advances are coming from spacecraft and satellite observations involving multimillion dollar funding. Ground-based telescopes and their auxiliary equipment represent smaller but still several million dollar investments. The 'man in the street', who must pay for these through taxes, may well grimace at such sums, but basic solar research has produced many practical applications, and no doubt will continue to do so in the future. For example, the discovery of absorption lines in the spectrum of sunlight led to the development of spectroscopy, the most powerful method of chemical analysis available today. The theory of the formation of absorption lines provided the basis for understanding the optical properties of paints, pigments and all opaque substances. Discoveries of the properties of matter at the high temperatures of the solar corona have helped in the development of nuclear reactors (and *vice versa*). Nobel prizewinner Hannes Alfvén developed concepts of the interaction of magnetic fields with highly ionized (see p. 40) and hence electrically conducting gases (which form the vast bulk of the Universe) in attempts to explain solar phenomena. His ideas form the starting point in attacking the problem of the controlled fusion of hydrogen atoms, one of the hopes for the world's future power supply. Radio studies of the Earth's ionosphere in the 1930s led to radar in World War II. Developments from this led to the discovery of the emission of microwave radiation by the Sun's atmosphere, and of magnetic fields extending far into interplanetary space, and opened up the new subject of radioastronomy. In 1975 the International Civil Aviation Organization opted for future world aerial-navigation, a microwave scheme developed by solar astronomers as a direct application of radiotelescope technology. Again, the properties of specialized high-resolution antennae developed for solar and non-solar radioastronomy have been applied to pinpointing satellites and space probes, and to improving medical X-ray analysis.

Not often do we recognize that all life on Earth is completely dependent on the Sun. In our concern about the exhaustion of fossil fuels – all containing solar energy stored from bygone ages – we usually overlook the vast bulk of the world's energy which is due directly to solar irradiation. This includes heating of the Earth's surface and distillation of water from the oceans. Without this the land would be parched and temperatures would sink rapidly below polar extremes, extinguishing all life. The extra energy, mostly of fossil origin, that we use in machines is but a tiny fraction of our free, unappreciated endowment from the Sun.

The Sun may not always be so generous however. The great Ice Ages of the past, at least three within the past million years, and lesser fluctuations within historical times, suggest that the solar power output is modified from time to time in manners presently unknown and unpredictable. We can be very certain that a minor Ice Age such as occurred in the seventeenth century and which may well have been caused by a major reduction in sunspot activity (the 'Maunder minimum'), would so reduce world food production that a sizeable fraction of the world's population could not survive. In the final chapter we shall be considering the probable course of sunspot activity in the next 50 years and its implications, severe droughts, which may well produce similar results. In the meantime it will be vital for astronomers and solar-terrestrial physicists to study intensively the causes of the Sun's sporadic fluctuations and the mechanisms whereby these may influence weather.

1   The Sun as we see it

The Sun seems to be a white-hot disk with an extremely sharp edge, but we know that it is spherical in form. We often speak of its 'surface'. This can give the quite wrong impression that there is a sharp outer boundary. Later we shall see evidence that the Sun extends far outwards. It has no sharp boundary – the density falls off continuously from the centre to the 'surface', and out through the planetary system, to merge beyond into that of interstellar space. In a sense, the Earth is immersed in the Sun's outer atmosphere, but its density is extremely low here – less than 1% of a billion billionth* of that of our own atmosphere.

By the Sun's 'surface' we mean the very shallow layer known as the 'photosphere', seen in ordinary white light. Although the surface seems uniform, a telescope shows it to be strongly textured. The most obvious features are the sunspots, dark cool regions surrounded by intensely hot solar gases (Fig. 1).

The Sun changes daily. Usually there are sunspots present, but sunspot activity varies greatly from a minimum (sometimes zero) to a maximum and down to a minimum again in about 11 years. This is called the sunspot or solar cycle. The picture in Fig. 1 was obtained in 1957, near sunspot maximum.

Spots occur in groups which contain large spots (the biggest here is 40 000 km across, three times the diameter of the Earth) and minor ones, down to pores only 1000 to 2000 km in diameter. Sometimes a 'group' contains only a single spot. A small individual spot may last for a day or less, but the big groups and big spots may be present for 3 or 4 months.

The very existence of sunspots is intriguing. They should be heated quickly from the sides and disappear. They should never have formed – but they do form. Their behaviour is so strange that there is still argument between scientists as to why they are there at all.

The Sun is full of mysteries apart from sunspots. Rarely do we discover anything that we would have expected. This makes solar research exciting. There are sometimes gaping holes to be filled in the fundamental laws of science before we can explain what happens.

Sunspots claim our interest, too, because they and other types of solar disturbance associated with them can have substantial effects on the Earth. They may improve or interfere with radio transmission, produce geomagnetic storms and generate great auroral displays in polar regions. Many scientists are also satisfied that they affect the weather.

Near sunspot minimum, there are only a few occasional small spots, and sometimes none at all (Fig. 2). At all times another quite surprising phenomenon is very obvious – the brightness of the Sun's surface is much greater at the centre of the disk than near the outer part of the disk (the 'limb'). The Sun certainly doesn't look that way to the eye even when screened by cloud, fog or a dark glass, because the eye is not very good at detecting gradual changes. If you wish to try for yourself, **DO NOT** look directly at the Sun! This can quickly damage your eyes permanently. Ordinary sunglasses must not

---

* Billion is used throughout to mean one thousand million (1 000 000 000).

1

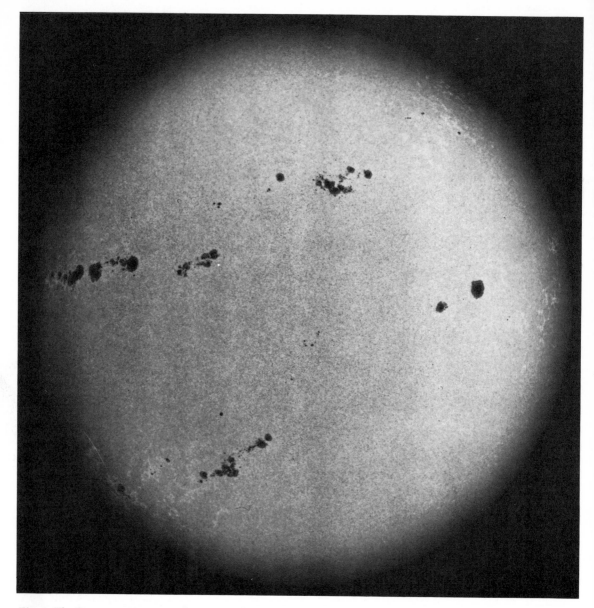

Fig. 1. *The Sun near sunspot maximum – 21 December 1957. The dark sunspots are aligned in two parallel zones. Also to be seen are fine, bright features known as faculae. Sometimes they are close to the sunspots, and in most cases they lie in the sunspot zones.*

be used either; they have a completely different function. You **must** use a proper electric-welder's eye-protection glass, which absorbs invisible but dangerous infrared radiation; or a photographic negative or smoked glass so dense that you can see the Sun quite comfortably. It will still appear uniform, but measurements show the centre to be 2.5 times brighter than the edge. In

Fig. 2 the contrast has been increased photographically, and the brightness variation across the surface is *very obvious*.

The reason for this variation is quite simple. Rays from the centre of the disk come vertically from the surface. Rays from near the edge travel along highly inclined paths. In both cases, light escapes only if it isn't reabsorbed. Since the shortest path from

Fig. 2. The Sun near sunspot minimum – 28 July 1954. No spot was present at that time. The brightness falls off from centre to edge, although this is not noticeable without photographs. No attempt should be made to look directly at the Sun to check on this unless through proper electric-welding glass, or very dense smoked glass or fogged photographic negative film, because of the danger of permanent damage to the eye.

any point to the surface is along the vertical, there is least absorption along vertical rays. Thus we can see deeper into the Sun near the centre than the edge of the disk. As temperature increases with depth, the Sun appears hotter or brighter at the centre.

The fall in temperature and pressure with height (Fig. 3) may be found by analysing the variations in brightness from centre to limb. This method can be used only for that part of the atmosphere from which the bulk of the Sun's radiation escapes, the photosphere. It is a very thin layer, only about 300 km deep, under 0.05% of the Sun's radius, and contains only about one-fifth of a billionth of its mass.

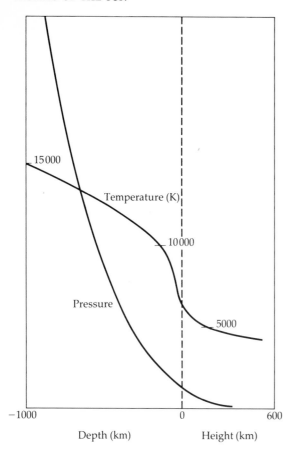

Fig. 3. Variations of temperature (K) and pressure near the Sun's surface. Zero height represents a particular level in the photosphere below which it is difficult to see. Higher levels can be observed in many different ways. Temperature and pressure both increase rapidly inwards.

unlike the descriptions given by the visual observers. The majority of Janssen's pictures exhibited large-scale distortions which he believed to be of solar origin. He was wrong but his contemporaries were not to know this and the subject remained confused. Today we know that the distortions are due to temperature variations in the Earth's atmosphere – such as cause stars to twinkle, or the loss of resolution when we look over a hot roadway.

Distortions of this type present by far the biggest single problem in making high-quality solar observations. Astronomers speak of the quality of atmospheric 'seeing', which is very good if the temperature of the air above and in the telescope is very uniform; and poor if it is very non-uniform. The difficulties for daytime observations are much greater than at night-time, when there is no direct heating of the ground. Best conditions usually occur early in the day, before the ground becomes too hot. They are invariably associated with air which is warmer than the ground. These conditions are usually associated with a high-pressure atmospheric system. The telescope itself also becomes heated by the Sun, giving rise to non-uniform heating of the air inside the telescope. Thus great care is needed in designing a high-quality solar telescope, choosing the best site, and selecting the best moments for making observations.

Far too little attention was given to these matters until the late 1950s. In the meantime, controversy continued about the nature, size and shapes of solar granules – and almost all other solar structures.

One of Janssen's granulation pictures, obtained close to the disk centre, was quite

The temperature drops with height in the photosphere from about 8500 to 4500K, but the average is 5770K. The reason for the drop is simple; the Sun loses heat by radiation and this must be replenished continuously by a flow from within. Since heat flows only from a hotter to a cooler region, the inside must be hotter than the outside.

Like willow leaves to one early visual observer, like rice grains to another, the surface of the Sun has an elegant granular structure when seen at high resolution (Figs 4 and 5). The story of granulation observations illustrates the typical difficulties involved in solar research. Around 1877, Pierre Janssen in France began to obtain the first high-quality solar photographs. In one of his best (taken in 1883), individual granules appear bright, have average diameters about 2000 km, and are surrounded by dark lanes. They are quite outstanding (Figs 6 and 7). The individual granules were remarkably well defined, of very uniform sizes, and separated by much narrower dark lanes than in any of his other pictures. As recently as 1953, K. O. Kiepenheuer (then the leading solar astronomer in Europe) declared this to be the best photograph of granulation ever obtained, but some curious things are associated with it. There is a remarkable alignment of granules parallel to the side of the page (Fig. 7), which continues to the limits of the original pictures, ten times as far as in the extract here. This alignment has never been observed on any other granulation photograph near the disk centre.

There is also a puzzle associated with the narrowness of the dark lanes. The ability of a high-quality telescope to resolve fine structure, under perfect seeing conditions, is proportional to the diameter of the telescope lens or mirror, and Janssen's lens was too small to show features so narrow – his other good photographs (Fig. 4) agree closely with what would be expected.

It was not until about 1955 that French astronomer Audouin Dollfus, re-examining Janssen's original emulsion, found it covered with a pattern of fine cracks – the cause of all the trouble. Janssen himself had not examined the plate carefully enough to notice the defects which were produced almost certainly by large temperature differences between the processing baths. Such a minor oversight led Janssen, the leading solar photographer of his day, to produce three different types of picture of the solar surface. His contemporaries could only be confused as to its nature. Were all three real; or which was, if any?

All controversy about the nature of solar granulation was finally resolved by magnificent photographs obtained in 1957 with a telescope carried to a height of 24 km (80 000 feet) by a balloon. At this height, the effects of the rarefied atmosphere are greatly reduced and the shapes of the granules are resolved clearly.

In recent times the quality of ground-based telescopes also has been so improved that, with great patience, high-quality images are obtained at several observatories (Fig. 8). These observations, together with those obtained from balloons, now allow the properties and behaviour of granules to be studied in detail. Their remarkable velocity structure will be described in Chapter 4.

Individual granules turn out usually to be irregular polygons of widely differing sizes, separated by dark lanes. In some places can be seen exceedingly small bright points, the filigree, no more than 100 to 200 km across; these are about the smallest features observed on the Sun until now, although the cores of flares are thought to be much smaller still.

Although a granule may be larger than Britain, New South Wales or Texas, it usually disappears in under 10 minutes, and may have a violent birth and death (Fig. 9). What is the cause of these short-lived structures? Do they have a major effect on the behaviour of the Sun, or are they too mild to influence it significantly? Some of the answers appear in later chapters.

Now let us see a sunspot and its neighbourhood at high resolution (Fig. 10). Delicate fibrils stretching out radially from a dark core (the 'umbra') form a lighter

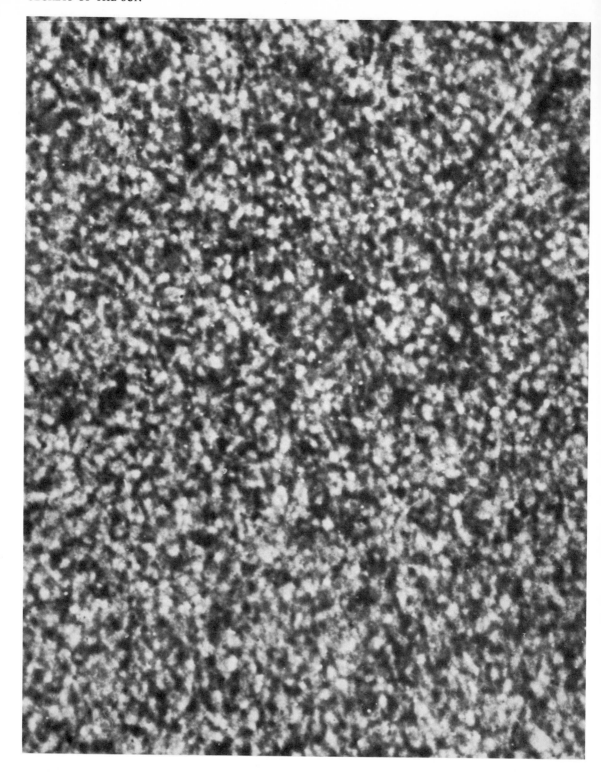

Fig. 4. *A high-resolution photograph of the Sun's surface obtained by Pierre Janssen in 1883. The structure is of individual bright granules surrounded by a dark background, commonly known as dark lanes. Granules are typically 2000 km in diameter in this picture, but smaller sizes are present also on modern photographs.*

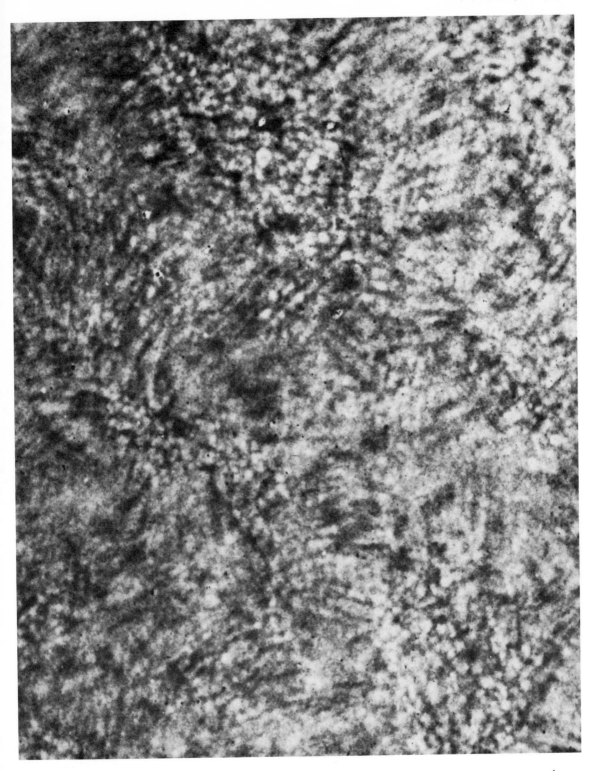

Fig. 5. *Most of Janssen's photographs showed strange features as here. He thought that these were present on the Sun, but in fact they are due to thermal non-uniformities in the Earth's atmosphere, or even in the telescope. Even now these defects of 'seeing' are the biggest obstacle to studying the Sun in detail.*

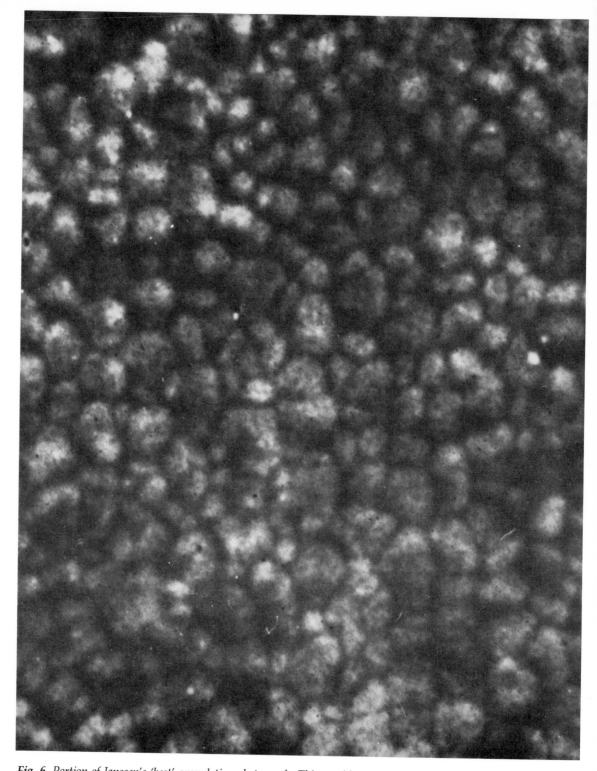

Fig. 6. Portion of Janssen's 'best' granulation photograph. This would seem to show the bright granules and dark lane structure extremely well, but has some curiosities of granule alignment which do not appear on modern photographs.

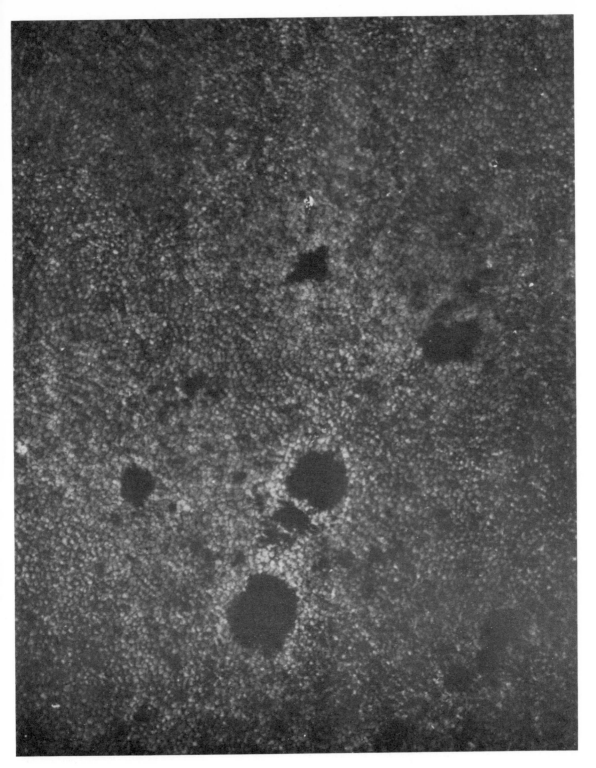

Fig. 7. *A large portion (lower magnification) of Janssen's best photograph (Fig. 6), showing the curious 'granule' alignment. The figure is tilted at about 30° from Fig. 6. The alignment is due to a pattern of fine cracks which remained unnoticed for over 60 years.*

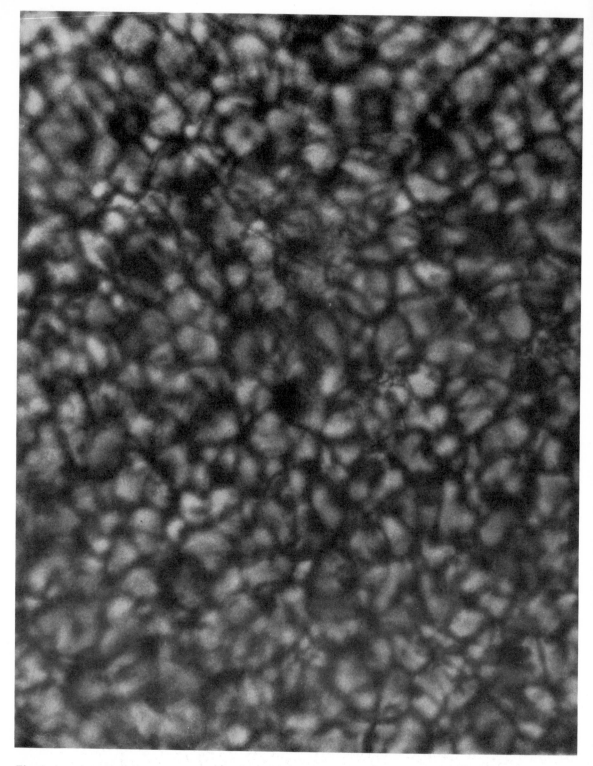

Fig. 8. *A modern granulation photograph, to the same scale as Fig. 6. Note the irregular shape of the granules, and the very small features present in places. Near the centre of the frame is a dark pore, a tiny feature like a miniature sunspot.*

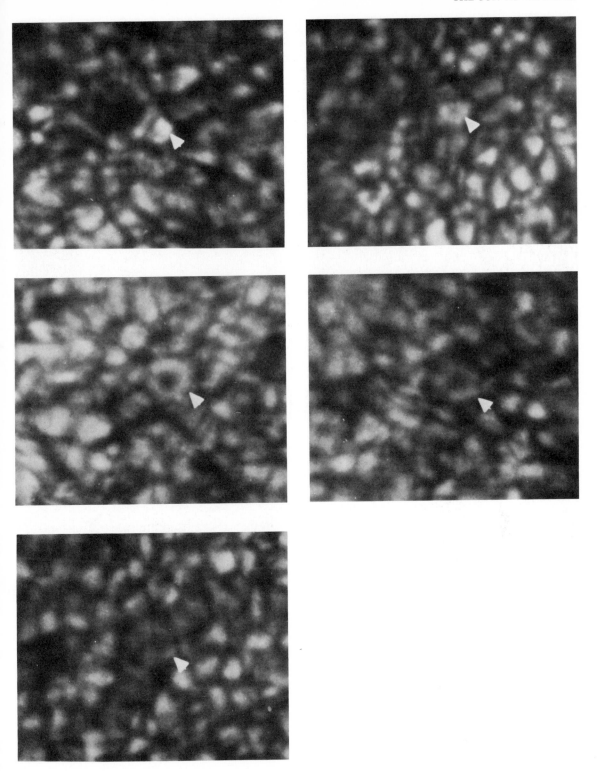

Fig. 9. *Occasionally granules explode, as shown here. The arrow points to a granule which expands and develops a dark centre in 8 minutes. Within 12 minutes, it has fragmented into four or five individual granules (times of the individual frames are 0, 4, 8, 10 and 12 minutes, widths of the frames 15 000 km).*

11

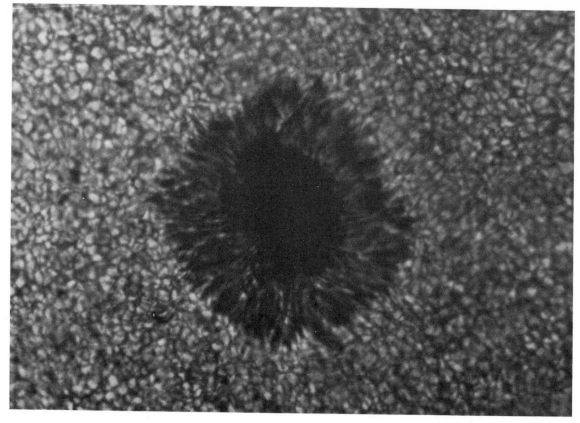

Fig. 10. *An isolated sunspot and the neighbouring granulation. The sunspot consists of a dark central 'umbra', and a lighter 'penumbra'. The penumbra is a mass of fibrils, some of which seem to cross over others. Near the outside we can see granules between gaps in the fibrils.*

'penumbra' in most sunspots. The umbra covers about half of the spot diameter and has a temperature typically 4000K, greatly below the average photospheric temperature of 5770K. Some sunspots are beautifully symmetrical, as here, but in others the penumbra and even the umbra can be very irregular.

The outside of the penumbra is fringed, and photospheric granules appear seemingly unchanged between projecting fibrils. In this way we get the impression that the penumbra is a shallow structure, the outer parts (at least) lying on top of the photosphere.

The early photographs (e.g. Fig. 7) showed sunspots surrounded by a ring of abnormally bright granules, which was believed for over 50 years to be connected with heat rising from the sunspot and diverted elsewhere. Modern photographs show no such rings, which were probably due to a well-known failure of some emulsions to reproduce intensities faithfully in regions of sharp contrasts.

This is another example of the need to understand every aspect of the behaviour of the instrument or detector (including photographic emulsions, photocells and detector arrays) with which a scientific measurement or observation is made. Unless a telescope and its associated equipment are performing exactly as expected, it is exceedingly easy to misinterpret the observation. The consequence is often persistent confusion that can have adverse effects on the course of astrophysics for decades.

2 Inside the Sun

The Sun is made of gas; nothing but gas. Like all other stars, the Sun is evolving. About 5 billion years ago it formed from a large cloud containing almost pure hydrogen gas and dust, like that in Fig. 11. The cloud contracted under its own gravitational forces, losing gravitational energy in doing so and converting it to heat. This process is still going on in many parts of the Universe and can be studied observationally, but the precise details of the prehistory of our Sun cannot be recaptured, and the exact way in which the planets, including the Earth, were formed is still under debate.

Until about 60 years ago, it had been thought that the release of gravitational energy was the main source of the Sun's heat. It is easy to calculate the total amount of heat that could have been produced: enough to keep the Sun operating at its present rate for about 20 million years. But geological evidence shows that the Earth is about 4 *billion* years old, and the Sun is a little older. A much greater energy source must be available. This turns out to be the conversion of matter into energy, as in atomic power plants and nuclear weapons. Present power plants split heavy uranium atoms into lighter atoms whose combined masses are less than the initial uranium mass. The difference appears in other forms of energy, particularly heat; but the processes used in today's atomic power plants cannot operate in the Sun since there are too few heavy atoms present.

The effective mechanism is the combination or *fusion* of four hydrogen atoms to produce a single helium atom. This is also lighter than the four hydrogen atoms, and energy is liberated as a consequence. This happens in the Sun via two different sets of nuclear reactions which proceed at different rates depending on the temperature. Both reactions need high temperatures, 10 to 15 million degrees. Thus, as time goes on the hydrogen concentration of the Sun decreases (now about 92% of the total number of atoms) and the helium concentration increases (now nearly 8% of the atoms). The other elements are scarcely more than impurities. The Sun is effectively a hydrogen ball. The Sun still contains enough hydrogen fuel to maintain its present output for another 10 billion years.

Not all the energy liberated in the fusion of hydrogen atoms results in heat. 'Neutrinos' are also produced. These are fundamental particles having less than one-millionth the mass of a hydrogen atom. Their most interesting property is that they have no electric charge. Every atom consists of a massive, positively charged nucleus surrounded by one or more electrons. The electrons, which are very small, have negative charges, but atoms are neutral as the positive and negative charges are balanced. At high temperatures (above about 10 000K) electrons are knocked off the atoms, which are then said to be 'ionized' or 'ions', i.e. electrically charged. The vast bulk of the Sun is highly ionized, and the free electrons make it an extraordinarily good electrical conductor. Even so, every particle, electron or ion, makes frequent collisions as it is attracted or repelled by its charged neighbours.

The neutral neutrinos are quite unaffected by electrical forces and almost all escape readily from the Sun. From theory we can calculate the numbers of neutrinos to be

13

expected at the Earth. This is the first possibility of an observational check of what is happening inside the Sun. An occasional interaction *does* take place between neutrinos and atomic nuclei, and these may be detected by suitable equipment placed underground to protect it from stray radiation. It turns out that there are fewer neutrinos than expected. The explanation is not yet clear, but it is already impacting on our knowledge of energy production near the centre of the Sun and will probably yield a direct measure of the temperature there.

We have referred to temperature many times already without saying precisely what it is. In a gas with no regular motions, the atoms are still moving about at random. Temperature measures the mean kinetic energy of the atoms. The kinetic energy is half the mass of the atom multiplied by the square of its random velocity ($\frac{1}{2}mv^2$), so that for an atom of a given type, for example oxygen, the mean random or thermal velocity increases as the square root of the temperature. Increase the temperature by 100 times and the mean thermal velocity of these atoms increases by $\sqrt{100}$ times, i.e. by 10 times.

Solar gases are mainly hydrogen, which has a mean thermal velocity of nearly 130 kilometres per second (km/s) at one million K, about 3500 times the speed of a supersonic fighter aircraft. An electron has less than 1/1830 the mass of a hydrogen atom, and its thermal velocity is about $\sqrt{1830}$ times, or nearly 43 times greater, about 5500 km/s at the same million K temperature!

How does the heat produced reach the Sun's surface? There are three possibilities: conduction, convection, and radiation.

Conduction is the process by which heat flows through matter, as hotter atoms (or electrons) hit cooler adjacent ones and pass their heat on. It is involved, for example, when heat flows through any metal. Thermal conduction is negligible inside the Sun, but we shall see in Chapter 8 that it is important in the outer parts of the Sun's atmosphere.

Thermal convection is the process whereby hot gases carry heat as they move bodily from one part of the Sun to another. When operating, convection is most effective in transporting heat.

Radiant heat flow is the dominant transport mechanism throughout most of the Sun. Radiation is emitted when hot electrons collide with atoms. The radiation travels at the speed of light (300 000 km/s) until absorbed. Although in free space radiation would travel a distance equal to the Sun's radius in little more than 2 seconds, absorption is so strong inside the Sun that it takes 1 million to 2 million years to diffuse out. Thus, the light and warmth we receive were produced near the Sun's centre over a million years ago.

If radiation were the only mechanism carrying heat, the Sun should have a tranquil, featureless surface, but the presence of granulation and sunspots indicates that this is not so. We can guess that there are motions present, probably due to convection.

To study this further we need to know the variations of temperature and density with depth inside the Sun. To derive these we use the facts that (1) at every level the gas pressure just balances the force of gravity, and (2) the temperature varies with depth in just the right way to transport the heat which is radiated from the Sun's

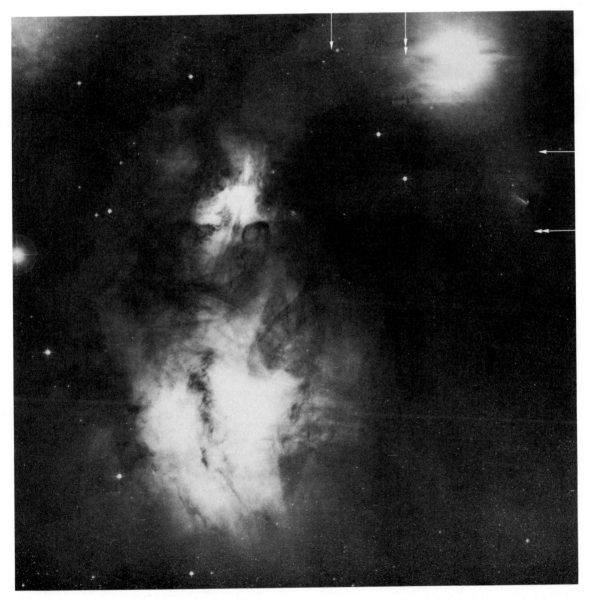

Fig. 11. *Dust clouds in the constellation Ophiuchus where solar-type stars are being formed. The darker regions indicated by arrows show the dust and its associated hydrogen gas. The extraordinary linear structures in the bright clouds are due to the all-pervading presence of magnetic fields. Myriads of background stars lie between the darker clouds.*

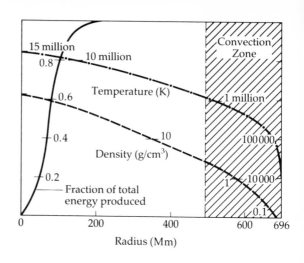

Fig. 12. *Temperature and density throughout the Sun, together with the fraction of its heat produced within a given radius. The centre of the Sun is at the left-hand edge, the surface at the right-hand edge. The convection zone, about 200 000 km deep, is hatched.*

surface. To solve the problem we proceed in stages, assuming first that heat is carried by radiation alone, and find the resulting temperature and density everywhere. Then it is necessary to test whether, under these conditions, convection would occur.

The test is rather simple. It is well known that gas cools on expanding – this is the process used, for example, in every kitchen refrigerator. If gas rises to a higher level where the pressure is lower, it will expand, and therefore cool. What happens next depends on whether the gas becomes cooler or warmer than its new environment. If cooler, it is denser than its surroundings and sinks harmlessly back to its original level; but if it is warmer, it is lighter than its surroundings and continues to rise. In the special case where the surroundings are at the same temperature reached by the rising gas, the region is said to have an *adiabatic temperature gradient*.

The convective process is remarkably effective, and limits the fall in temperature with height to the adiabatic temperature gradient almost wherever convection can operate. It applies equally to the Earth's atmosphere. It is solely responsible, for example, for the drop in temperature on going up a mountain, and for producing clouds and rain when moist air cools on rising.

It turns out that radiative heat transport operates alone most of the way from the centre of the Sun to the surface, but in a zone whose depth is rather uncertain theoretically, from the photosphere down to between say 100 000 and 225 000 km, convection takes over and the temperature gradient becomes adiabatic (Fig. 12). Chapter

4 shows how recent observations are starting to tell us about the depth of the convection zone, around 200 000 km. In the meantime, we can be impressed with the central temperature of the Sun – 15 million K – and its density, 160 times that of water; and we may be intrigued that energy production is largely confined to a small volume where the temperature exceeds 10 million K, within 140 000 km of the Sun's centre.

The convection zone is exceedingly important for many of the phenomena occurring at the Sun's surface. It is unfortunate that very little information on convection under strange stellar conditions comes from any type of laboratory experiment conducted so far. The critical difference between Sun and laboratory is the great range of density (100 000 times or more) between the bottom and top of the solar convection zone, whereas laboratory experiments have been effectively at uniform density. A big density range might well alter the pattern of convection dramatically, but frankly we do not know. To make progress, theoreticians are currently calculating the behaviour of individual gas particles under solar conditions so as to infer the convection pattern as a whole.

In the meantime we can try to build up an idea of what might be happening by considering some naturally occurring convection patterns. First we shall see what the solar convection zone is *not* like. Under very uniform conditions and with relatively small gas velocities, a very regular convection pattern develops both in laboratory experiments and in the Earth's atmosphere when, for example, the air is heated uniformly from below. Convection takes place then in vertical cells, known as Bénard cells, in which lighter, warmer air rises in the centres of the cells and denser, cooler air falls around the boundaries. The well-known altocumulus (Plates I and II) and cirrocumulus clouds are the result of this type of convection.

At much greater velocities, a regular pattern no longer occurs. The gas motions become unstable and turbulence sets in, with the development of turbulent eddies. The conditions for turbulence are known well enough, and it is clear that convection in the Sun is turbulent, but of this process, we know very little that is applicable to astrophysics. Some common examples of turbulence can be seen in a rising plume of cigarette smoke in very still air, when after a short distance the steady plume becomes unstable and its behaviour quite irregular; or in the smoke from a chimney stack, particularly in a breeze, where the turbulent eddies can be seen and their changes followed. A large-scale example of turbulent flow and eddy formation comes from the Voyager observations of Jupiter (Plate III). Like the Sun, Jupiter is almost pure hydrogen, although its surface features are clouds of solid ammonia flakes. Sets of eddies, the largest ever observed, are churned between zones of latitude rotating at different rates and between various other structures in relative motion. The eddies do not retain their identities, but gas at the outside of an eddy is drawn out in long tongues, perhaps to form further eddies.

We can only guess that some analogous process occurs inside the solar convection zone, but is more complicated in at least two additional ways. First, we *see* eddy formation on Jupiter only in a two-dimensional plane, whereas inside the Sun eddies are constantly formed in three dimensions. Secondly, the eddies observed in the Earth's atmosphere and in the laboratory are effectively at uniform density, whereas the density falls off by a huge factor in the solar convection zone.

Finally, in the photosphere itself, radiative heat transfer takes over once more and a most important and unexpected collision process occurs – electrons become attached to hydrogen atoms to form *negative* hydrogen ions. When this happens, these ions emit light – the vast bulk of sunlight and solar radiation is emitted in this way.

3 The Sun's curious rotation

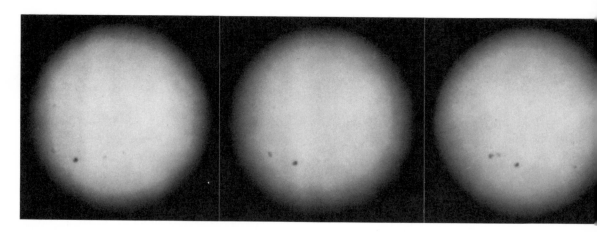

Fig. 13. *Probably in December 1609, just prior to the first application of the telescope to the study of the Sun, Johannes Fabricius observed sunspots using a technique similar to that of the pinhole camera, which uses no focusing lens or mirror to form an image. In this modern repetition of his observations, we can see the high quality of the images obtainable. Photographs at daily intervals demonstrate solar rotation. Photography being a recent development, early recordings were all by drawings.*

From the time of the discovery of sunspots around 1610, it has been known that sunspots are carried across the disk of the Sun from edge to edge in 13 to 14 days. Galileo's telescopes enabled him to observe solar rotation with great ease, but even without a telescope the German astronomer Johannes Fabricius could note the daily change in sunspot positions using a camera obscura. A modern reproduction of the types of image that Fabricius would have seen is shown in Fig. 13, where the daily progression of sunspots is obvious. Overall, the Sun rotates once in about 27 days, which happens to coincide with the periodic rate of fluctuation in the Earth's magnetic field. We shall see later that there is good reason for this coincidence.

Since the Sun rotates, there is an axis of rotation (the polar axis) and an equator. We can use the same system of latitude and longitude as on Earth for specifying location on the Sun. The Sun's axis is tilted a little more than 7° from a right angle (a 'right angle' = 90°) to the plane of the Earth's orbit (the Earth's axis is tilted at 23.5° from this direction), so that at times we may just see the Sun's north pole, and 6 months later its south pole.

It has been known since 1863 that the rate of rotation of the sunspots depends on their latitude – on their distance from the equator. No such thing could happen on the *solid* Earth, but a stranger on the Moon might well use clouds or storms to measure the Earth's rate of rotation. As these move faster than the solid Earth at high latitudes (the 'roaring forties'), the Moon observer would find the rate of rotation of the atmosphere *increasing* with latitude.

In the example just given, the clouds or storms are tracers; so are the sunspots. The

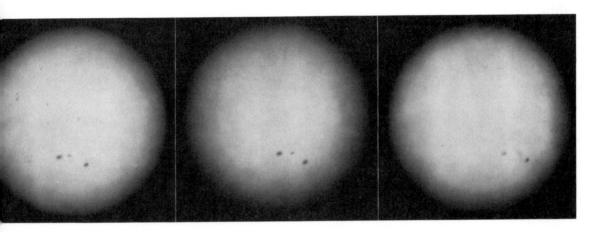

intriguing thing is that they show that the Sun rotates fastest at the equator, just the opposite from our own atmosphere (Fig. 14).

One of the problems in using sunspots as tracers is that they are rarely found further than about 30° from the equator, virtually never beyond 40°, so that they can tell us only about rotation at low latitudes. Is there any way of checking on the sunspot results, and of extending the law of rotation to higher latitudes? There are indeed other types of feature such as filaments and surface magnetic fields, to be described in later chapters, that can be used, not quite as well defined and as accurately as sunspots, but covering a greater latitude range, to about 60°. They tell very much the same story, with the rate of rotation falling off even more rapidly with increasing latitude. The variation of rotation rate with latitude is known as *differential rotation*, and can occur only because the Sun is gaseous. Sunspot drawings made by the German astronomer Johannes Hevelius in 1642–44, and even earlier drawings by Scheiner, can be used to

study the differential rotation of about 350 years ago. There was scarcely any difference then from now, so that the rate of rotation derived from sunspots has been virtually unchanged over many centuries.

A quite different and more direct method of measuring rotation is based on spectroscopy. In principle this method has been available for a long time, but only recently has modern technology been able to provide the accuracy needed.

To appreciate it, we need first to discuss light waves and spectral lines. Light is a wave propagating at extreme speed, 300 000 km or 186 000 miles per second (Fig. 15a). Just as with waves on water, the light waves have peaks and troughs, the separation between consecutive peaks or troughs being the wavelength, λ (Fig. 15a). In light, the wavelengths are very, very short, perhaps 20 000 to a centimetre. The colour we see depends on wavelength, reds being longest of the visible wavelengths, and blues the shortest with yellows and greens in between. It is merely a matter of evolutionary luck that

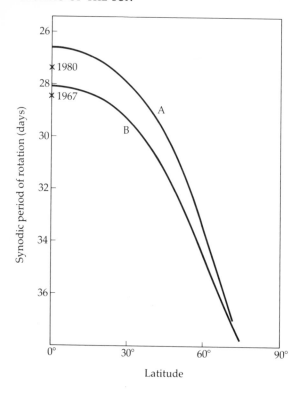

Fig. 14. *The rate of rotation of the Sun is greatest at its equator. Curve A shows the synodic rate (i.e. the rate as measured from the Earth) derived from tracers such as sunspots, prominences and other phenomena of magnetic origin. Curve B shows the rate of rotation of the gases as measured by Doppler shifts. The gas rotates more slowly than the tracers. Curve B was the average Doppler shift around 1970, but the rate had been increasing at the equator from 1967 to 1980 (crosses). Differential rotation of this sort and variations with time can occur only because the Sun is gaseous and not solid.*

most of our eyes can distinguish these wavelengths by colour; a few people are completely colour-blind and are unable to detect any colour variation whatsoever. There are also longer invisible wavelengths which lie in the infrared, with radio waves longer still. And at shorter, invisible wavelengths there are the ultraviolet, X-rays and gamma-rays.

When any one type of atom is excited, for example by collisions with electrons, it emits light of sharply defined wavelengths which are characteristic of the element. These can be separated in a spectrograph. This is an instrument which works like a camera. Light enters through a narrow slit, and a lens

or mirror forms a sharp image of it, except that the image is displaced sideways by an amount which varies with the wavelength. If the light has a set of sharply defined wavelengths, we find a set of sharp, separate images of the slit, each appearing as a narrow line. In the case of *white* light, we find instead a continuous patch of light varying in wavelength from one side to the other (Plate IV). The dispersed light is known as the spectrum, and the individual wavelengths as spectral lines.

The same wavelengths are absorbed by the atom when light falls on it, so that atoms in the Sun's atmosphere produce dark absorption lines by which they can be identified (Plate V). This is the basis for the chemical analysis of the Sun's atmosphere.

The stronger absorption lines are given simple names. These include the very strong and red C or Hα line, and the blue F or Hβ line (both due to hydrogen), the close yellow D_1 and D_2 lines (both due to sodium) and the close green b_1, b_2 and b_4 lines (all due to magnesium). There are other well-known lines such as H and K, due to calcium, well off the blue end of Plate V. Other spectral lines can be identified simply by their wavelengths. Often we give first the chemical symbol for the atom and its degree of ionization. The degree of ionization is 1 plus the number of electrons that have been removed from the atom, and is always written in Roman numerals. Thus C III means a carbon atom with two electrons missing, and Fe X an iron atom from which nine electrons have been removed. This is followed by the approximate wavelength in Ångström units (1 Ångström unit = 1 Å = 10^{-10} metres). Thus we can speak of the Hβ

(a)

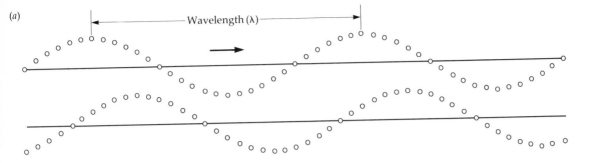

(b)

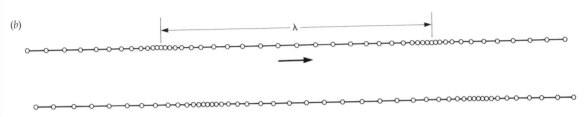

Fig. 15(a). A transverse wave (e.g. light) propagating from left to right. Below it we see the appearance one-sixth of a period later. The wavelength λ, is the distance between consecutive peaks or troughs. (b) A longitudinal wave (e.g. sound) in which the displacement is in the direction of propagation. This results in places where the medium (e.g. air or solar gases) is more concentrated, i.e. denser, separated by more rarefied regions.

line as H I 4861, b$_2$ as Mg I 5172 and K as Ca II 3933.

In 1842 the Austrian physicist Christian Doppler discovered that when an object emitting waves moves towards or away from the observer, the wavelength becomes shorter or longer: 'the Doppler effect' (Fig. 16). This is the cause of the well-known change in pitch of an engine whistle as a train passes at high speed, since the pitch (frequency) of a wave increases as its wavelength decreases. It is also used in the radar detection of cars exceeding speed limits. We make great use of this principle in astronomy, as the Doppler effect is independent of the distance of the source. If we can identify a spectral line, and the wavelength differs from what we measure in the laboratory, the velocity of the source towards or away from us is given immediately by the change in wavelength. We can even do better than this – by subtle analysis, we can study *differences* in velocity between gases at different heights in the Sun's atmosphere.

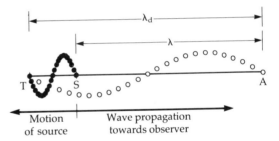

Fig. 16. Origin of the Doppler shift. The velocities of waves depend only on the medium in which they propagate, and not on the velocities of the source. Here a transverse oscillator (e.g. a light source) moves away from the observer, going from S to T in one period, the time required to make one oscillation (filled circles). Open circles show the wave propagating towards the observer on the same scale as Fig. 15(a). At a given instant, the wave which had started from S reaches A, just as far from S as if the source had been at rest; the distance SA is the wavelength (λ). One period later the source is at T and the corresponding wave is just leaving T. The actual Doppler-shifted wavelength is λ$_d$, which is longer than λ by the distance travelled by the source in one period, ST. This may be checked by comparison with Fig. 15(a). We call this a 'red-shift'.

Similarly, the wavelength would be decreased if the source were moving towards the observer. We call this a 'blue-shift'.

21

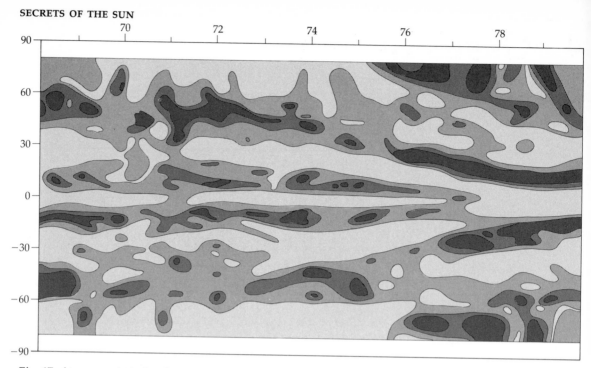

Fig. 17. *At any one latitude, the velocity of rotation fluctuates systematically about the mean rate during the cycle. This is known as the torsional oscillation of the Sun. Here regions rotating faster than normal are shown darkest; those slower than normal are lightest. One can trace the way in which faster rotation drifts from poles down to the equator in 22 years, during which two fast-rotation waves would have started from each pole. When the rate of fast rotation passes latitudes of about ±30°, a new sunspot cycle commences.*

Doppler shifts don't let us see transverse velocities, motions perpendicular to the line of sight. If we want to study vertical motions on the Sun, we can use Doppler techniques at the centre of the disk because the line of sight is vertical there. For horizontal motions we may use the same techniques towards the limb, where the line of sight becomes more nearly parallel to the surface.

The simplest way of measuring rotation by Doppler shift is to find the difference in the wavelengths of a given spectral line at opposite edges of the Sun, or at equal distances from the central meridian (Plate VI). The first precision measurements in 1967 found that, near the equator, the rotation rate of the gases was some 6% slower than the tracer rate, indicating that the tracers were moving faster than the gases in which they were embedded. The rotation rate was kept under close watch in the subsequent 13 years, during which it increased by about 4%, still below the rate of the tracers (Fig. 14).

What does all this mean? Every sunspot is connected with solar magnetic fields. It seems that the regions where the magnetic fields originate, well below the surface, are rotating faster than the photospheric gases, pulling the fields forward through the photosphere. Relative motion between fields and non-magnetic gas, no matter how it originates, turns out to be an essential part of the theory of the solar cycle, to be described in Chapter 7.

A most exciting result on solar rotation has come from a recent demonstration from Doppler-shift measurements that, in addition to its regular differential rotation, there is a superimposed torsional oscillation (Fig. 17). In this, the polar caps first rotate slightly faster than average, then slower, the time required for an oscillation being 11 years. The oscillation then drifts to the equator in 22 years. These periods are just those associated with the sunspot cycle and the solar magnetic cycle. The connection of the torsional oscillations and the cycle is also described in Chapter 7.

4 Gases in motion

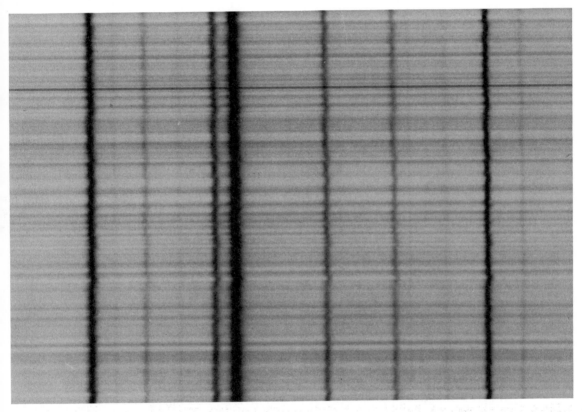

Fig. 18. A high-resolution spectrum showing absorption lines of various strengths (vertical). The direction of dispersion, or wavelength variation, is horizontal. The lines are not straight but irregular (particularly noticeable on the fine lines) due to motions in the granules. The broader lines and the continuum show where the surface is bright (the granule centres) or dark (the granule boundaries). The brighter regions are rising (blue-shifted), and the darker falling (red-shifted).

Almost all sorts of motions are present at the surface, except whirlpools – which is rather surprising in view of current ideas of the convection zone. The absence of surface eddies shows that the top layers of this zone, at least, are not as turbulent as expected. What we see are convection cells, which hold out tantalizing promises to tell of conditions inside the convection zone. Through their interactions with the magnetic field, they also dictate the structure of the Sun's atmosphere. There are flows of gas from near the equator towards the poles, and there are oscillations and wave motions which are providing means of probing not only the Sun's atmosphere, but also its hidden interior. Waves are also believed to heat the upper atmosphere to high temperatures. There are curious gas-flows in sunspots and in other features that we have not yet discussed. Some of these motions will be described in later chapters.

Convective motions were discovered first in the granules, described in Chapter 1. If an image of the Sun is focused onto the slit of a spectrograph, the spectrum lines show wiggles due to Doppler shifts produced by the granules (Fig. 18).

The geometrical pattern of motion can be found by a quite different technique. While the spectral lines are narrow in wavelength, they still have appreciable widths. Optical filters have been constructed which allow light to be transmitted from only extremely narrow parts of the spectrum, such as a *portion* of a spectral line. If we take two filtergrams, one on each side of the spectrum line where granule motions cause opposite intensity fluctuations, and subtract one picture from the other, there remains a pattern showing only the motions. It turns out that the brighter granule centres are moving upwards and the boundaries, about 30% darker, downwards (Fig. 19). The typical difference in velocity is about 0.4 km/s, say 1000 miles per hour. Such speeds, high by terrestrial standards, are quite small on the Sun.

The granules have quite simple explanations. The top of the convection zone lies in the low photosphere, and we can see the tops of various convection cells. The hotter parts move upwards, cool by radiating to outer space, and flow horizontally towards the granule boundaries. There, now cooler and therefore denser, they sink back into the Sun. If of the simple Bénard-cell type described in Chapter 2, the granules would

have long lives and be of regular patterns, but this is not so. The lifetimes, some 10 minutes, are about the time for gas to flow just once around the cell, while the cells are of irregular shapes and sizes. We are dealing with 'non-stationary convection'. The diameters are a few times the depth of the zone (a few hundred kilometres) where, because of appropriate temperatures, hydrogen atoms are partly ionized. It is this rather special property that extends the convection zone so high and is responsible for the granules. Unfortunately they can provide no guide as to the nature of convection at greater depths.

Convection patterns can also be seen on a much larger scale, about 30 000 km across. These are the supergranules (Fig. 20). They have lives of about a day and horizontal velocities of about 0.4 km/s towards the boundaries. The vertical velocities are too small to be measurable, 0.01 km/s at most. As with granules, gas flows around a supergranule cell about once during its lifetime. By contrast, there is scarcely any variation in brightness from centre to boundary. Nevertheless we seem once again to be dealing with non-stationary convection, now in a zone about 8000 to 10 000 km deep where helium, the only other element present in any substantial amount, is partially ionized. Again the supergranules can give no guide as to the nature of convection in the bulk of the convection zone.

Between granules and supergranules there is a distribution of sizes on which convection occurs. This can be demonstrated (Fig. 21) by the effects in producing small magnetic network cells (see Chapter 7), or

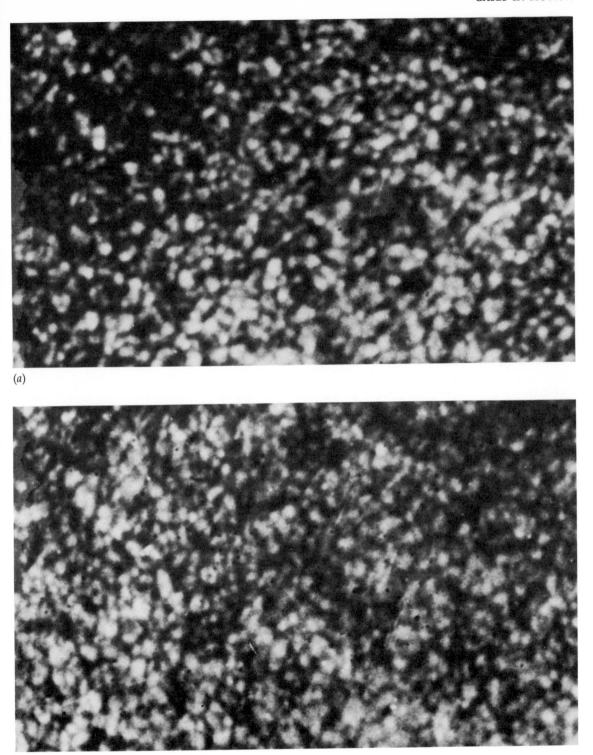

(a)

(b)

Fig. 19.(a) *Normal photograph of granulation.* (b) *Display of vertical velocities in the granules; upward motions appear bright, downward dark. It is interesting to make detailed comparisons, granule by granule, between brightness and velocity.*

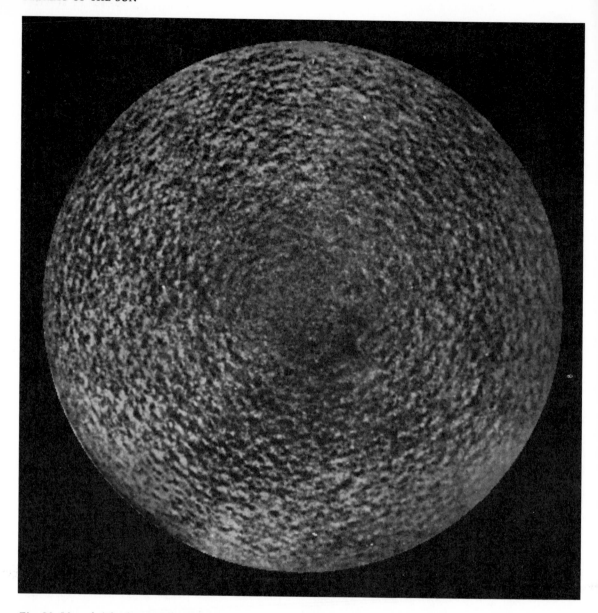

Fig. 20. *Line-of-sight distribution showing the supergranulation pattern. Regions with components of motion towards the observer are shown dark, and away from the observer, bright. Since the pattern is absent at the centre of the disk, vertical supergranule velocities are very small there. The obvious patterns well away from disk centre are due to horizontal motions up to about 0.4 km/s in cells typically 30 000 km across.*

directly just as for the supergranules – indeed on the same 'velocitygrams' as show the supergranules. The intermediate sizes of convection (Fig. 22) are weaker than the supergranule and the other motion patterns. Their lives are intermediate between granule and supergranule; they average about 2 hours.

There are systematic gas motions of about 20 m/s away from the equator between latitudes 10° to at least 70°. These do not seem to vary significantly during the solar cycle. They will be shown in Chapter 7 to have profound effects on the distribution of the magnetic field over the Sun and they are vital to the theory of the solar cycle. Long-lasting surface flows of gas away from

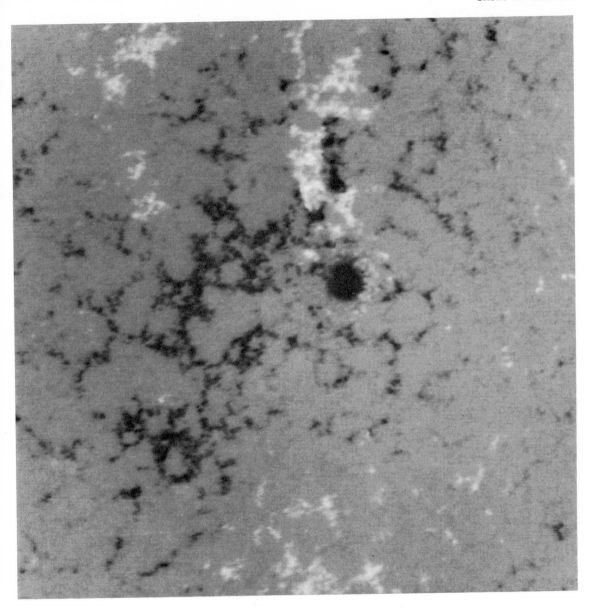

Fig. 21. A magnetogram showing magnetic networks of varying sizes. While the majority of the networks have diameters about 30 000 km, regions of heavy magnetic packing show much smaller networks. How many can you find? Since the larger networks are known to be at the boundaries of supergranules, to which the supergranule flow is directed, the inference is that convection also occurs on scales intermediate between those of supergranules and granules. They are invisible where magnetic fields are not closely packed, so that this picture alone is insufficient to prove their existence elsewhere as normal features.

the equator must be matched, of course, by a subsurface flow towards the equator so as to maintain continuity; otherwise there would be a huge pile-up of gas near the poles and the Sun would quickly become elongated. As the density increases with depth, however, the velocity of the subsurface counterflow must be much smaller than the observed surface velocity.

It is almost unthinkable that a mass of gas should be at rest. For example, 'oscillations' occur whenever an object is disturbed from a condition of rest. There are just two essentials for oscillations: (1) all forces acting on the object at rest must be exactly in balance; and (2) if the object is displaced, the forces must act so as to return it to its original position. Since the displacement of any particle disturbs its neighbours, which push back on the original particle and also hand their own disturbances on to *their* neighbours, a wave is propagated.

A star is a mass of gas held in equilibrium under the force of gravitational attraction opposed by gas pressure. Conditions are just right for oscillations and many stars oscillate violently by swelling and contracting, some with periods as long as a year, others with periods of the order of a few days or less, and maximum radius commonly several times the minimum.

The Sun oscillates too, in many different ways, but the amplitudes are small; the total range of movement may be only of the order of 25 km. Not much, you might think, by comparison with a radius of about 700 000 km, but enough for us to measure.

The account of oscillations given below is rather technical, and may be difficult for some readers to follow, but it will be worthwhile making the attempt. Fundamentally, waves are responsible for the existence of the corona and thereby influence the whole of interplanetary space including the Earth's upper atmosphere. In addition, the theory of these waves is very precise, quite an unusual feature in astrophysics, and yields methods of studying the interior of the Sun which would not be available otherwise. If the description of waves proves to be too difficult, the reader may skip to Chapter 5.

Many of the Sun's oscillations are due to sound waves, in which the displacement is not transverse (Fig. 15a) as with waves on the ocean or light waves, but longitudinal (Fig. 15b). The speed of propagation is nearly equal to the mean thermal velocity of the atoms and hence is proportional to the square root of the temperature, as explained on p. 14. On Earth the velocity of sound is about 330 m/s. In the solar photosphere it is some 20 to 25 times greater because the temperature is much higher and the atoms (mainly hydrogen) are much lighter.

Now let us follow a sound wave as it travels (propagates) through a region of varying temperature, increasing vertically downwards, say. For waves of any type, a wavefront is a surface on which all points are in phase and, in all cases we shall meet, the wave propagates at right angles to the wavefront. Suppose the sound wave is propagating downward at some angle to the vertical. In Fig. 23, a wavefront, W, is shown at a particular time, and two points, A and B, on it. B is lower than A and therefore hotter, so the speed of sound is greater at B. After a brief interval, the wave will have travelled further from B than from A, and the wavefront will tilt. The direction of propagation changes, and continues to change until the wave turns almost vertically upwards. It undergoes total internal reflection! This process is exactly like the reflection of sound that occurs in the Earth's atmosphere wherever the temperature increases with height, e.g. when the ground and nearby air cool at night, or over water, creating a temperature increase in the air just above. It is for such a reason that sound travels so well at night; sound waves which would otherwise escape are reflected back towards the ground.

We may now study sound waves in the

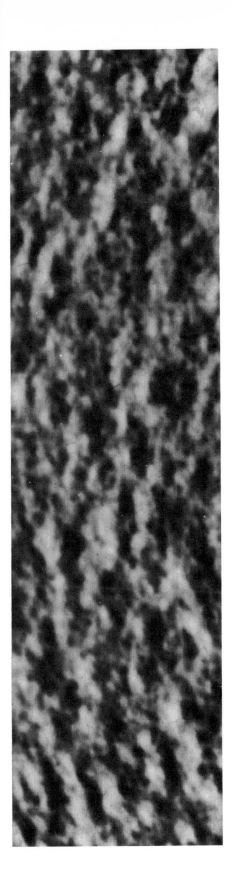

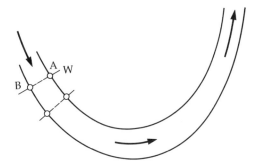

Fig. 22. *A larger-scale picture of horizontal velocities in supergranules when averaged over 40 minutes to eliminate short-period oscillations. Smaller-scale structures can also be seen, intermediate in size between granules and supergranules. They are thought to be responsible for intermediate-scale magnetic networks, as in Fig. 21.*

Fig. 23. *Propagation of a sound wave inside the Sun, where the temperature increases inwards (downwards). W is a wavefront at a given time, but B is deeper and therefore hotter than A, a point higher on the same wavefront. Thus the velocity of sound at B is greater than at A, and after a short while the new position of the wavefront is tilted from its former orientation. The wave follows the path shown, being totally reflected upwards. The steeper the initial direction at W, the deeper the wave goes before reflection.*

Sun's convection zone, where they must be generated copiously by turbulence. The first evidence of oscillations of the solar surface was obtained just over 20 years ago, when a team led by Robert Leighton, from the California Institute of Technology, discovered oscillations having periods of about 5 minutes (10 000 times as long as typical sound waves on Earth). It required intensive study by many astronomers before the nature of the oscillations became clear, due primarily to theoretical analysis by Roger Ulrich, also of California, and its observational confirmation by Germany's F.-L. Deubner.

The detailed mechanism is fascinating. Depending on frequency and direction of propagation at some reasonably high level, waves generated in the convection zone and

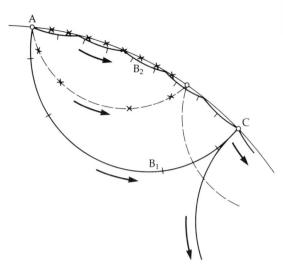

proceeding downwards will be totally reflected by the increase in temperature, as we have just seen.

Waves propagating upwards can also be reflected downwards near the top of the convection zone for a quite different reason. The fall in density with height becomes very rapid as the temperature drops in the photosphere. The sound wave coming up from below strikes an almost abrupt boundary above which the atmosphere becomes almost a vacuum. Suppose the gas density were uniform above this. Since the wave amplitude is the same immediately above the boundary as below, but the density differences are extreme, only a minute fraction of the wave's energy could be carried upwards above the boundary. The density continues to fall rapidly with height, however, and in consequence no wave energy passes at all. Reflection is again total! The wave bounces up and down between the two reflection levels.

The atmosphere above the upper level oscillates as a whole, since it moves up and down because of the pressure fluctuations from below. This is called an *evanescent* wave. If the temperature decreased continuously with height, no energy could flow. Later we shall see that, higher in the atmosphere where the temperature increases with height, energy can flow upwards once more; this modifies our conclusions, but not very much. The disturbances observed by Leighton and others have been mainly evanescent waves, since the upper reflecting layer is a little below the photosphere.

We have seen that turbulence in the convection zone generates sound waves which propagate between upper and lower

Fig. 24. The two ways in which a sound wave may leave point A on the Sun's surface and arrive in phase at C: (1) after total internal reflection along path B_1 inside the Sun; (2) by hugging the surface, path B_2, which it cannot penetrate because of the savage continual drop in density. When two waves such as B_1 and B_2 arrive at C in phase (i.e. differing by an integral number of wavelengths, here 5 as against 9 as shown by the cross markers), the waves reinforce and become strong. Otherwise they interfere and are of negligible amplitude.

The dashed lines show how similar reinforcement may take place at a higher frequency, but only along different wave trajectories (for clarity, the surface wave is not shown).

reflecting levels, but only some of these can be observed. To understand this, consider a reflection point, A, on the upper boundary, Fig. 24. The reflected wave traverses the path AB_1C to C, where it is reflected once again at the upper boundary. There is one other path, and only one, by which a wave can travel from A to C; this (AB_2C) makes grazing incidence reflections, and therefore skirts just under the upper reflection level, which it cannot penetrate. Moreover it travels more slowly than along AB_1C since the temperature and velocity of sound are lower on average in the upper layers. If the distance is just right for the two waves to arrive exactly in phase (not necessarily an integral number of wavelengths as shown in Fig. 24 by cross-ticks), they will reinforce. In most cases they tend to destroy one another. Let us see what this implies.

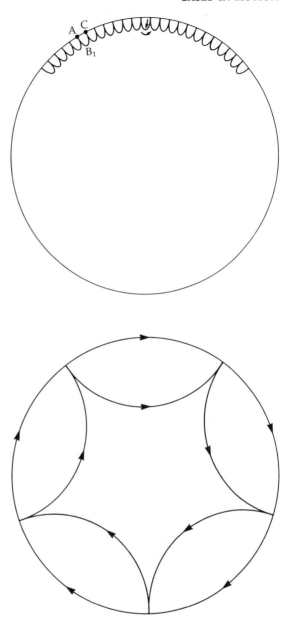

Fig. 25. The type of wave system (i.e. small distance AC) in the usual 5-minute solar oscillations. The two waves which reinforce are AB₁C and AC, which travels directly under the surface. Waves can propagate in either direction as shown by arrows. As observed, these waves appear to have different properties because of solar rotation which extends one and contracts the other depending on direction. Presumably the wave system disintegrates eventually because of irregularities.

To be exactly in phase, travel times along the two paths AB₁C and AB₂C must *differ* by an exact number of periods, nT, e.g. $n = 4$. For some higher frequencies (or shorter wavelengths of the sound wave; see Fig. 25) this number stays the same provided the wave AB₁C propagates further from the vertical in the upper part of the atmosphere, i.e. the wave returns to the surface closer to A (the broken curve in Fig. 24 on which wavelengths are marked by crosses). A plot of frequency against the distance AC should be a continuous line. For a different n, a different line is found. Ulrich's predicted shapes and positions of these lines are shown in Plate VII.

To study these predictions observationally, Deubner measured the vertical velocity distribution spectroscopically for many hours along part of the Sun's equatorial diameter under conditions where distances AC were typically one or so thousandths of the Sun's circumference. His results agreed remarkably with theory. Small residual differences have already been used to improve models of the convection zone whose base, for example, turns out to be some 200 000 km below the photosphere.

Observations of this type can also be used to measure the rate of rotation at different depths. The waves pictured in Fig. 25 can proceed in either direction around the Sun. But the Sun itself is rotating, so that the point C turns closer to or further from A after traversing its appropriate path depending on the direction of wave propagation. In other words, the measured distance AC depends on direction and on the actual velocity of rotation. As path AB₁C occurs at some appreciable depth inside the

Fig. 26. The wave situation where only a small number of reflections is involved around the whole Sun. In this case the waves must all be in phase after completing a circuit of the Sun, adding a major restriction to the number of waves that can exist. Ulrich's continuous curves (Plate VII) then break up into discrete points.

Sun, the difference between the two directions of propagation measures the rotation at some average depth. As different waves penetrate to different depths, the rotation rate can be measured as a function of depth. Latest results by John Harvey and Thomas Duvall of Kitt Peak National Observatory in the USA provide an anticlimax – they find no variation as large as 1% down to 40 000 km! Quite inexplicably at present, their rate of rotation is the same as that of the sunspots, not of the gas as one might expect.

It is also possible to study wave modes with only a few nodes around the whole Sun, in which case the distance AC can be a large fraction of the diameter. There is then no need to focus an image of the Sun. A condition to be satisfied in this case is that there must be an integral number of reflection points around the Sun; otherwise the waves destroy rather than reinforce one another (Fig. 26). In this case, the lines of Plate VII would not be continuous but rather the loci of discrete points.

First demonstrated by George Isaak and his collaborators, impressive results of this type have been obtained by a joint French-American expedition (Fig. 27) which carried out observations from the South Pole in 1979/80. During summer, they obtained uninterrupted observations over 5 days, or 7 days if two brief interruptions were allowed.

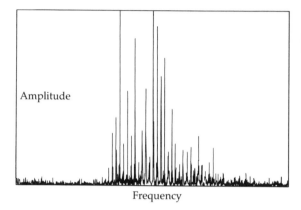

Fig. 27. Discrete frequencies observed by Grec, Fossat and Pomerantz originating as described in Fig. 26. With observations over a long-enough time interval, the rate of rotation can be measured deep inside the Sun. Initial results by Isaak and his collaborators indicate that the core of the Sun rotates much faster than the surface.

They found a sequence of extremely well-defined frequencies, as we would expect. Observations of somewhat longer duration by Isaak's team have been interpreted as demonstrating that the core of the Sun is rotating two to nine times more rapidly than the surface. The implications pose interesting problems for the future.

5 Those notorious sunspots

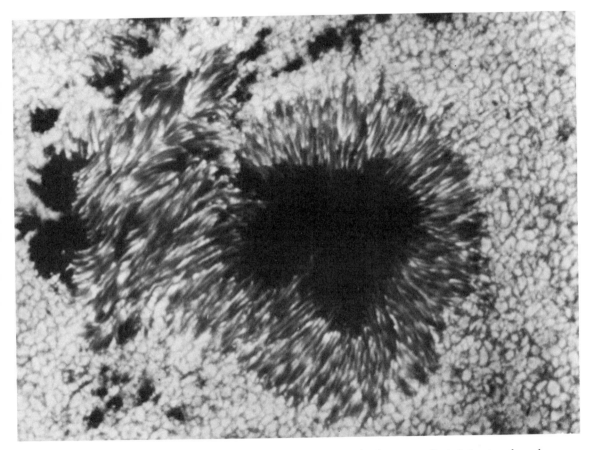

Fig. 28. *A large sunspot (the leader) and smaller followers. The penumbra has a complicated structure here, by contrast with that of Fig. 10. The penumbral fibrils are of two types, dark and bright. The latter have brighter heads near their inner ends. Some tiny pores can be seen to the bottom left, and near the top.*

Sunspots are impressive. Except during eclipses when coronal structures and prominences can be seen, the biggest sunspots are the only solar features visible to the unaided eye. A modest spot of diameter 10 000 km (1/140 that of the Sun) blocks or diverts over 20 000 times more energy than falls on the whole Earth's surface in the same time!

Sunspots occur in groups of from one to as many as a hundred, of which two are usually much bigger than the others. They are aligned approximately east–west. The first of the larger spots to rotate around the eastern limb is known as the leader (Fig. 28), the second as the follower. The group may last for only a day or less if very small, or for weeks to months if very large, during which time it may cross the disk by solar rotation several times. The smallest spots ('pores') are only a few times as large as granules, say 2000 km across. The largest have diameters up to 20 times greater.

A spot group develops rapidly to its maximum extent in perhaps one-tenth of its life, decaying much more slowly. The processes of growth and decay go on together. The smaller spots disappear first, and the follower before the leader, so that it is not uncommon to find the leader remaining as an isolated spot (e.g. see Fig. 10).

The dark centre of the sunspot, the umbra, has an intensity typically about 15% of the photosphere and a temperature of some 4000K. The umbra occupies usually about one-half of the sunspot diameter. Outside this is a fainter ring, the penumbra. Not all sunspots have penumbras; they are rarely present on the smaller spots, and even around larger ones the penumbra is sometimes incomplete.

What causes sunspots? How can they survive, embedded in the much hotter photosphere? The clue comes from the spectrum (Plate VIII). We have seen already that excited atoms emit spectral lines of unique wavelengths. If the atoms are in a magnetic field, each spectral line is split by the 'Zeeman effect' into components of differing polarizations and slightly different wavelengths whose separation yields the field strength. All sunspots are found to have strong magnetic fields of about 3000 gauss (about 10 000 times the Earth's magnetic field). In most cases the field is of one polarity in a single spot, but there are exceptions; when these occur, we can expect fireworks! Each group is bipolar; the leader

and follower are of opposite magnetic polarities. Usually the minor spots have the same polarity as the closer major spot. It is clear that sunspots and large magnetic fields are intimately linked.

The key property of magnetic fields, for sunspots and many other solar phenomena, is that they can move through good conductors only extremely slowly. Any relative motion between field and conductor induces electric currents, and forces between these and the field make their *relative* motion negligible (the same laws are used in the design of all electric generators and motors). It would take about 10 billion years for fields to escape from the centre of the Sun, and a still longer time to escape from or cut through the gases nearer the surface. During this time, the magnetic fields exert a pressure on the gas which works just as ordinary gas pressure, except that the lines of force also resist stretching. It is just as if they were elastic strings, in tension.

We are now ready to explain the dark sunspot. In the outer layers of the convection zone, heat is carried upwards by convective motions *and* by radiation. In the strong sunspot field, a convection loop cannot be completed because gas cannot flow across the field; heat flows almost by radiation alone, and therefore at a lower rate. Since less heat is carried upwards inside the field, the region appears dark at the surface; hence the sunspots.

There is an immediate consequence of the presence of strong magnetic fields in sunspots. Here the gas pressure and magnetic pressure inside the sunspot combine to balance the gas pressure outside, but a 3000 gauss field has a magnetic

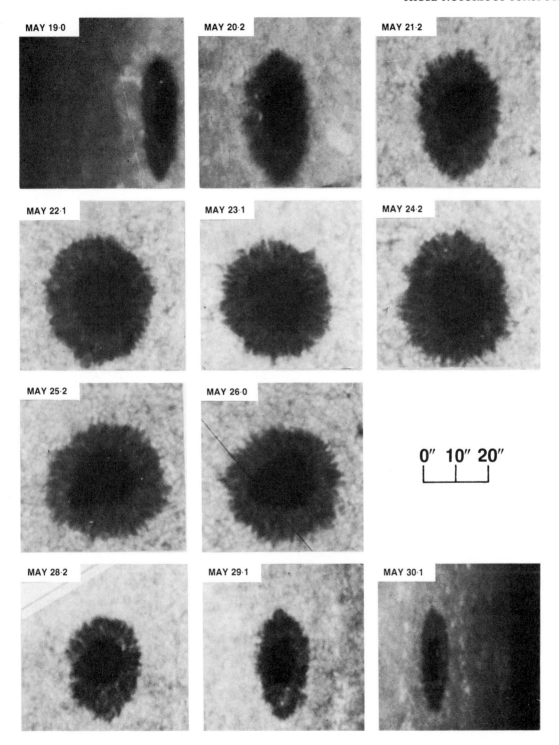

Fig. 29. *Changing appearances of a stable, symmetrical sunspot as it crossed the disk. The foreshortening of the penumbra near the limb is quite different on the limb and disk sides, the disk side appearing much the narrower. This effect, discovered by Wilson in 1769, is due to the umbra being depressed so that the penumbra is tilted. The changes from centre to limb enable the depth of the Wilson depression to be measured, and the curvature of the penumbra in a radial direction to be found.*

35

Fig. 30. Cross-section of an average, stable sunspot. The height scale is ten times the horizontal scale.

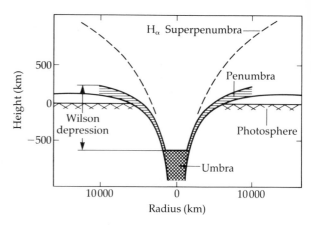

pressure of 36 000 pascals (Pa) alone, more than double the photospheric gas pressure, so that it seems as if equilibrium is impossible. Even worse; for where field lines are curved outwards, as in spots, magnetic tension adds further outward forces. Is there any way of resolving this?

An intriguing discovery by the Scottish astronomer Wilson in 1769 was that, when near the limb, the penumbra is much broader on the side nearer to the limb than on that further from the limb. This is clear proof that the umbra is lower than the penumbra and that the penumbra tilts down into the spot. The height involved is not easy to establish accurately and may well vary from one spot to another, but seems to average about 1050 km. A simple comparison shows that the external gas pressure at this depth is 350 000 Pa, which would crush the internal sunspot forces consisting of the gas pressure and magnetic field combined.

As in most cases, the solution is rather simple. At the outside edge of the penumbra there is no magnetic field until about 180 km *above* the photosphere level $z = 0$, virtually irrespective of the size of the spot; the base of the penumbra there cannot be lower than this. It is probably some 250 km thick, whence the umbra lies only about 620 km below $z = 0$ (Fig. 30). Finally, the umbral magnetic field is measured using an absorption line, the light from which originates higher than the umbral base, i.e. about 450 km below $z = 0$. Here the pressures are in balance with the magnetic forces to within the limits of uncertainty. The ratio of the internal to the external gas pressure is low, about 0.2, and all is well.

There is an elegant fine structure in sunspots. The penumbra is divided into more or less radial, brighter and darker filaments whose widths are down to 200 km, less than half the diameter of the smaller granules. Yet they have much longer lives, some 2 hours. At times it is possible to see fibrils crossing one another at substantial angles, suggesting strongly that the magnetic field direction may vary or rotate with height. The umbra is packed with a granulation-type structure known as 'umbral dots', also finer than photospheric granulation and of longer life, typically 15 to 30 minutes. The dots are much hotter than their umbral surroundings, about 6200K, even hotter than the photosphere. Why? We don't know. No sooner do we solve one puzzle about sunspots than others appear.

An intriguing flow of gas, the 'Evershed flow' (Fig. 31), occurs outwards along the dark penumbral fibrils at a velocity of over 6 km/s, about equal to or a little greater than the velocity of sound. The cause is unknown, but it shows that the penumbra is not in hydrostatic equilibrium. No flow is found in the bright fibrils.

The nature and origin of the penumbra are still uncertain. For some time it had been thought that each fibril represented a discrete tube of magnetic force, but it seems more likely that the field is fairly uniform. The individual fibrils are probably due to the way in which gas from an irregular source flows along the lines of force and not across them, preserving the source structure.

Plate 1. *Terrestrial altocumulus clouds just after sunset. The individual cloudlets are at the tops of rising, warmer convection elements. Between, where the clouds are broken or missing, the air is cooler and flowing downwards. The Sun pillar, a shaft of light rising vertically above the sunken Sun, is due to reflection from flat snowflakes aligned almost parallel to the Earth's surface. This shows that any turbulent motions present are so gentle that they scarcely disturb the orientation of the snowflakes in the clouds.*

Plate II. *Altocumulus clouds seen from above. While the individual cloudlets indicate the tops of rising columns of warmer air, they show considerable internal structure. Although the pattern of altocumulus clouds is formed in the same way as Bénard cells, the details of motion in the cloudlets themselves are complicated by the liberation of heat of condensation when cloud-drops form, and by heating by sunlight and subsequent cooling by radiation from the clouds. Surprisingly little is known of the detailed mechanism of this type of cloud.*

Plate III. *Although Jupiter consists mainly of hydrogen, its clouds consist of solid ammonia flakes. This Voyager 2 picture shows eddy formation near the great Red Spot, where opposing surface motions set up huge systems of eddies, possibly resembling more closely than any other we know the types of turbulent eddies that may be present in the solar convection zone.*

Plate V. *Spectrum of sunlight. Dark lines occur because selected colours are absorbed by chemical elements in the Sun's atmosphere. For example the Hα and Hβ lines are due to hydrogen, the D lines to sodium and the b lines to magnesium.*

Plate IV. *(Top) Continuous white-light spectrum with no spectral line. (Centre) Emission lines from excited atoms. (Bottom) Appearance of absorption lines superimposed on the spectrum when the atoms lie between the white-light source and the observer.*

Plate VI. *A diffraction-grating spectrum formed when an image of the Sun is allowed to fall onto the slit of the spectrograph. Top and bottom represent spectra of east and west solar limbs. The different colours arise because a diffraction grating produces overlapping spectra, normally removed by a suitable filter. The very fine lines are due to absorption by gases in the Earth's atmosphere; the broader lines are solar. The tilt between solar and terrestrial lines shows Doppler-shifts due to solar rotation (it is these that are measured to produce curves like B in Fig. 14). The broad solar lines are curved, not straight as might be expected. The lines are first shifted to longer wavelengths (i.e. to the red) by solar gravitation; secondly, the hotter rising gas in the centres of granules is brighter than the cooler descending gas near their edges so that, near the centre of the disk, the weighted Doppler effect displaces the line to shorter wavelengths. This does not occur at the limbs, because the motions there are vertical and at right angles to the line of sight.*

Plate VII. (Right) Ulrich's predicted graph of the number of oscillations in unit distance across the Sun's surface, plotted against wave frequency, for waves differing by an integral number of wavelengths from 1 to 4. (Above) Harvey and Duvall's latest observed graph corresponding to Ulrich's prediction. The right-hand side corresponds to Ulrich's graph, the left-hand side is to facilitate study of the rotation rate with depth. Differences of up to 20 wavelengths between the two paths can be identified.

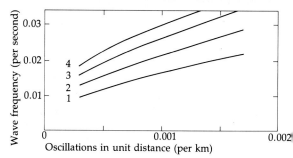

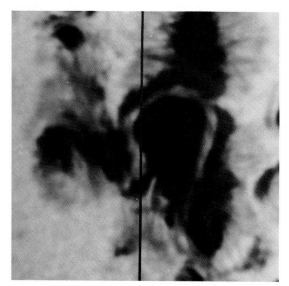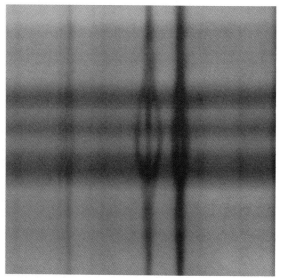

Plate VIII. A sunspot (left) and portion of its spectrum (right). The position of the spectrograph slit is shown by the dark straight line on the sunspot picture. Where the slit crosses an umbra, the spectral lines are split by the 'Zeeman effect'. The Fraunhofer line in the centre of the spectrum shows the most common type in which a single line is split into three, called a triplet. The amount of splitting is a measure of the magnetic field strength, and the relative intensities of the central and outer lines give the direction of the lines of force. The dark horizontal bands in the spectrum occur where the slit crosses various umbras.

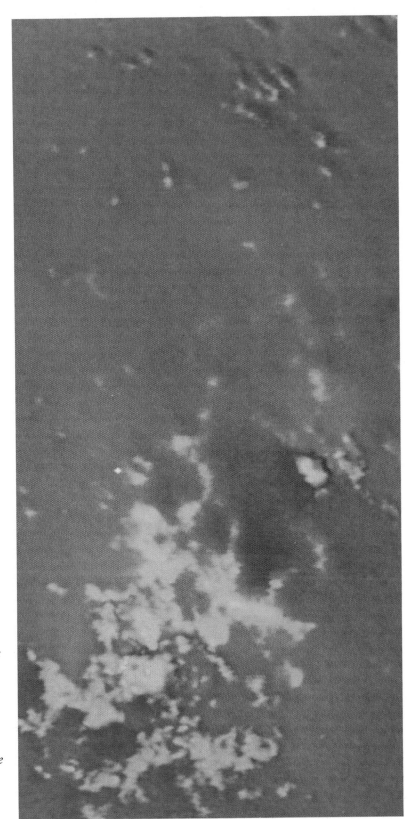

Plate IX. *Magnetic canopies near a weak active region. Regions where there are magnetic fields at photospheric level are shown in pink or yellow depending on polarity. Areas are coloured green or blue where almost horizontal fields lie above non-magnetic photosphere; these are the magnetic canopies. Their bases are at various heights, quite low, typically 300 to 400 km between the large pink area near the bottom of the picture to the small pink area half-way up.*

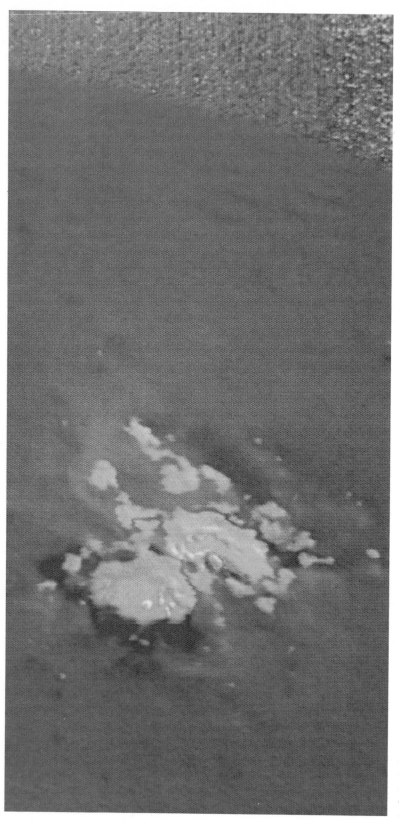

Plate X. *Magnetic canopies near a large sunspot (pink). Canopy heights (diffuse blue and green areas) are similar to those in Plate IX.*

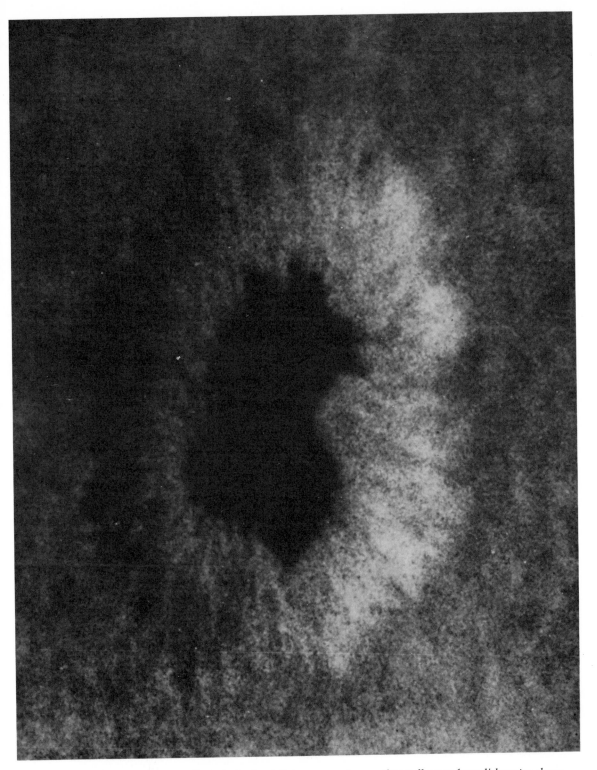

Fig. 31. *The velocity distribution (effectively horizontal) in a sunspot penumbra well away from disk centre shows the Evershed effect distinctly. Outward motions occur in the dark penumbral filaments at about 6 km/s, close to the velocity of sound.*

(a)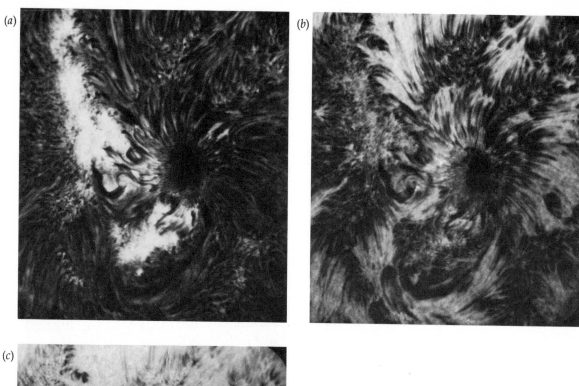

(b)

(c)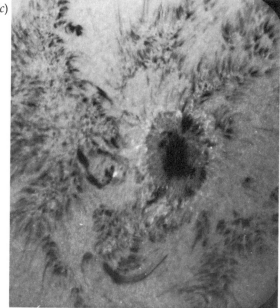

Fig. 32.(a) An isolated sunspot photographed at the centre of the strong chromospheric spectral line Hα. This shows the highest level observable at high resolution by ground-based telescopes. The umbra and penumbra are still detectable (compare with (c)). Surrounding these is a large set of almost radial fibrils seen only in Hα, the superpenumbra. A bright plage region marks where there are strong concentrations of magnetic field in the photosphere below.
(b) By changing to a wavelength at 0.5 Å from the Hα line centre, where the absorption coefficient is lower, we can see deeper, but still higher than in white light.
(c) At 0.75 Å from the centre of Hα, the sunspot has an appearance more like that in white light. We may notice a large number of tiny bright features, 'Ellerman bombs', or 'moustaches' from their appearance in spectra, fringing the penumbra.

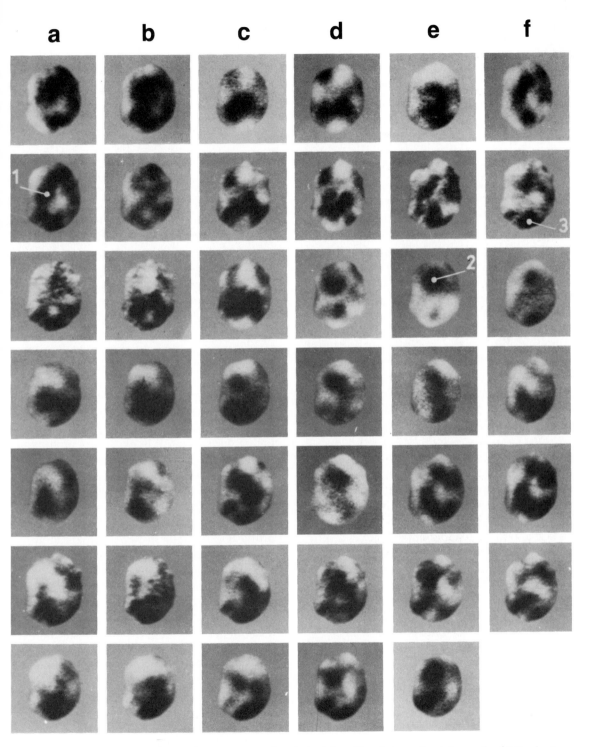

Fig. 33. *Vertical velocity distribution in a sunspot umbra derived from Hα photographs at inervals of 0.5 minutes along each row and 3 minutes down each column. Notice the tendency for the pattern of motion to repeat after 3 minutes, the pattern breaking down after a few oscillations. The points numbered show repetitions over several oscillations; see if you can find others.*

Sunspots can also be studied at higher levels using light from a strong absorption line. This absorbs all the light from the lower levels. The only light which can escape in the absorption line comes from higher in the atmosphere, the typical level depending on the line concerned (it is not always easy to know what the heights are). The most beautiful views are those in the strong Hα line of hydrogen (Fig. 32). The appearance varies depending on whether the light is from the centre of the line, where absorption is strongest and the level seen is highest (Fig. 32a), or from an intermediate position (b, c).

It has sometimes been suggested that the suppression of convection by sunspot fields, first proposed by Ludwig Biermann in wartime Germany, would be inadequate to explain their blackness. Other mechanisms would be needed, particularly the loss of energy by mechanical waves flowing out of sunspots. It is possible to test this using the technique described for showing velocity patterns in granules and supergranules, now with a filter transmitting two narrow passbands centred 0.25 Å into each wing of the strong Hα line. The umbra turns out to be bouncing up and down violently with periods of the order of 3 minutes and amplitudes 1 km/s or greater. These oscillations (Fig. 33) have nothing to do with the umbral dots, which are on a finer scale and whose origin remains mysterious. Limits can be placed on the energy carried away by waves propagating upwards from these oscillations; they account for only a few hundredths or thousandths of the sunspot's cooling.

Another interesting set of waves is found propagating outwards over the penumbra (Fig. 34). The displacement is perpendicular to the magnetic field, and the wave propagates along the field. Known as Alfvén waves, they are like the waves that travel along a stretched string or wire when tapped, as when a piano is played, or a violin. Tension pulls the displaced string back, and it then oscillates. In exactly the same way, tension in the lines of magnetic force, which acts to resist stretching, pulls the disturbed field back and sets a wave travelling along the field. Alfvén waves are very important for heating the Sun's outer atmosphere, and we shall meet them again in later chapters. In the case of penumbral waves, the energy carried amounts at most to one-two hundredth of that involved in cooling sunspots. All in all, losses of sunspot energy by mechanical waves are negligible; Biermann's mechanism is undoubtedly valid.

The creation of a spot group is quite clearly due to a well-defined subsurface tube of magnetic flux floating to the surface, cutting it eventually in two approximately circular sections, the leader and follower sunspots. It is impossible to envisage a tube of magnetic force subjected to the turbulent buffeting of the convection zone and not undergoing twisting and tangling. In consequence, we must expect the submerged tube to have detailed structure rather like a wire or rope which is wound from a large number of individual strands or fibres, a concept due to the Australian astronomer John Piddington. Observation of spot-group formation and growth has helped considerably in clarifying the properties of the flux rope. For example, Fig. 35 shows a curious granule alignment between major growing spots. In this case, several

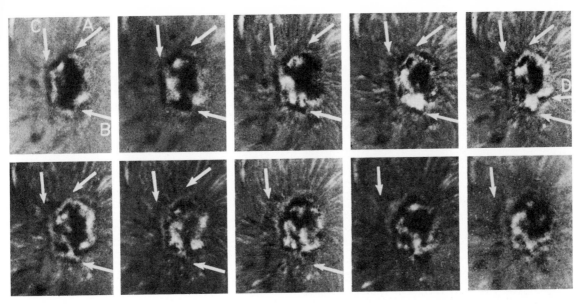

Fig. 34. *The vertical velocity pattern derived from Hα photographs of sunspot penumbras often shows a set of waves propagating outwards. Here, on velocity pictures derived at 0.5-minute intervals, arrows show successive positions of wavefronts. The velocity of propagation is typically 20 km/s.*

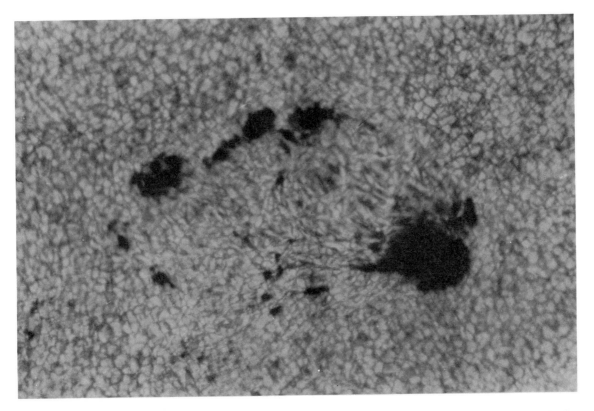

Fig. 35. *During the growth of sunspots, horizontal magnetic flux fibres rise through the photosphere and modify the granule pattern between sunspots. Here we see a striking example of this behaviour.*

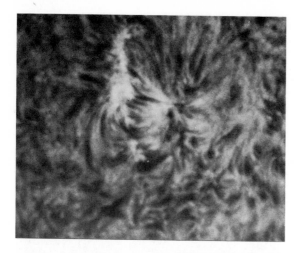

Fig. 36. *Arch filaments, observed in Hα, marking the emerging of flux fibres to higher levels than in Fig. 35. Arch filaments may appear first before the development of sunspots, and are usually present throughout the growth stage of a spot group. In these photographs, the arch filaments run more or less east–west through the centre of activity. The three pictures were obtained over a period of about 24 h.*

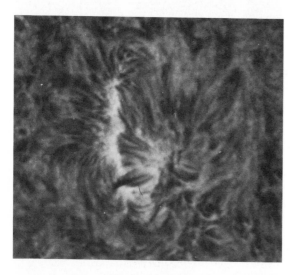

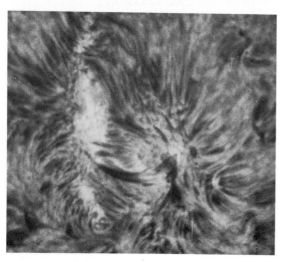

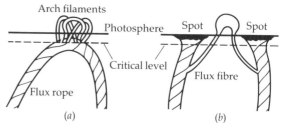

Fig. 37.(a) *Cross-section of the development of arch filaments. A twisted flux rope floats upwards to a critical level where the gas pressure is low enough for individual fibres to expand upwards through the photosphere into the chromosphere. In (b), a single flux fibre is shown after the whole rope has broken through the photosphere to form a pair of sunspots with arch filaments between them. It is the growth of these systems that causes the distortions of granulation seen in Fig. 35.*

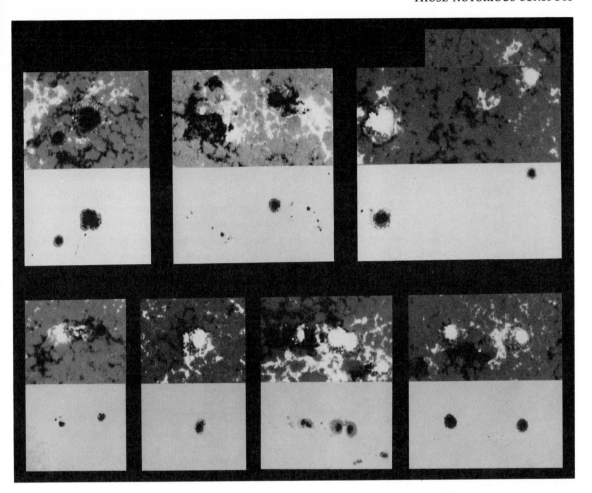

Fig. 38. *Sunspots decay by the unravelling of the twisted flux fibres. These then drift away from the spots, appearing as small magnetic elements clustered around them. In the examples seen here, the magnetic fields appear above, the sunspots below. While the intensity varies continuously across a sunspot, only a few intensity contours are shown here because of the method used for recording. In all cases, tiny elements of both polarities lie very close to a sunspot, but those of the same polarity as the spot outnumber those of opposite polarity by about two to one. This is because the majority of fibres develop kinks, each passing up through the surface twice and down once. Further from the spots lie the well-known unipolar plage fields which outline supergranule cells.*

individual flux strands rise through the surface, the magnetic tension in each being sufficient to keep the strands taut and to align the granules while rising between them. The ends of the strands form tiny spots or pores which migrate to join bigger ones.

At a higher level in the atmosphere, Hα observations show another view of this process. Between the growing sunspots is a set of arches (Fig. 36), each representing a flux fibre or strand which has just risen

through the surface. While individual fibres disperse, the system may last for a day or so while the whole flux tube floats up. A diagrammatic view of this process is shown in Fig. 37.

After the sunspots have formed, an intriguing phenomenon is discovered whereby the spot becomes surrounded by tiny, discrete elements of magnetic flux which move outwards (Fig. 38). Evidently the sunspot flux-tube is unravelling and the individual fibres are being carried away by

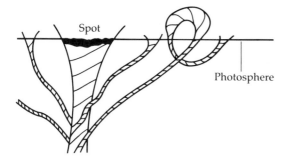

Fig. 39. Development of a kink in a twisted fibre.

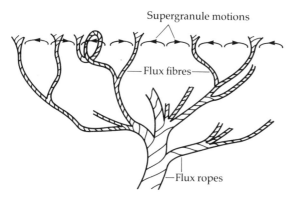

Fig. 40. Section of a rope below a decaying spot. Individual fibres unravel from the twisted flux rope and are carried away by horizontal motions. In the final stages of fragmentation, the original rope splits into hundreds or thousands of individual fibres or clusters. A few of these are shown in transit in the central parts of supergranules, although most have been carried to near supergranule boundaries by horizontal motions. There they remain, held between opposing horizontal flows, until the supergranule breaks down after about a day. They are then carried to the boundaries of the new supergranules.

motions in the surrounding gases. In a younger group where leader and follower spots are present, the space between may be filled with fibres of almost random polarities, whereas around an isolated spot (an old leader) the distribution is more uniform. In such cases, it is clear that some elements have apparently the wrong polarity. Piddington has offered a neat explanation of this; if any fibre or rope, no matter of what material, is twisted too tightly, it develops kinks (try it for example on the cord from a venetian blind). The fibres leaving the sunspot may well become unstable in this way and develop kinks (Fig. 39) which, on reaching the surface, produce tiny magnetic elements of opposite polarities.

The fibres are carried away from sunspots by supergranule flow (Fig. 40), at least in part, but the rate of migration of the magnetic elements seems too fast for this to account in full for the outward flow. One suggestion is that the penumbra acts as a blanket to reduce the loss of heat from below and to raise the temperature of the subsurface zone, thus enhancing the horizontal convective motions near the sunspot. In any case, the migration of tiny magnetic elements from the spot represents the mechanism of sunspot decay, which commences even during the process of growth.

The migrating magnetic elements are carried to the nearest supergranule boundary where they meet gases flowing in the opposite direction. There they stay, squeezed between opposing horizontal flows, to form a strong concentration of large numbers of discrete, magnetic elements known as a 'plage' after (goodness knows why) the French word for 'beach'. With the breakdown of the supergranules after a typical life of a day, the plage elements move around to new supergranule boundaries and the distribution of magnetic field near the spot group becomes more complicated. This region, the seat of many solar phenomena, is known as an 'active region'. As time goes on, the active region itself is dispersed by the repeated process of supergranule breakdown and reformation.

6 Magnetic fields outside sunspots and what they do

The sunspot cycle dominates the behaviour of everything that is of magnetic origin on the Sun and in interplanetary space. The first thing we may notice, apart from the sunspots themselves, is the great difference between the distributions of magnetic field near the sunspot minimum and maximum (Figs 41 and 42).

Near sunspot minimum, when there may be no spot present on the Sun for months, the magnetic field is concentrated in little points or 'elements' scattered apparently at random over the Sun, and of random polarities except at high latitudes (which are usually unipolar). However, we know from simultaneous observations of the magnetic field and surface velocities that the *locations* of the magnetic elements are far from random. They are transported by horizontal convective motions to near the boundaries of supergranules, where opposing flows meet.

As sunspot activity develops, the magnetic flux disperses from sunspots in the form of small, point-like concentrations as described at the end of the previous chapter, and lodges near the supergranule boundaries. An interesting phenomenon now converts the random or mixed-polarity distribution of sunspot minimum to the highly geometric unipolar distribution of sunspot maximum, in which large, skew areas of almost single polarity slant symmetrically in the northern and southern hemispheres. No matter where they occur on the Sun, the supergranules either become distorted into quite new shapes after about a day, or break down and are replaced by new ones. Whichever happens, the magnetic elements are carried to the new locations of the supergranule boundaries. On average,

they will be taken about a supergranule radius each day, about 15 megametres (Mm; 1 Mm = 1 million metres), but in a random direction (this is known as a 'random walk', a process well studied mathematically). Were this the only factor to be considered, magnetic elements liberated near a given point would be redistributed symmetrically about the origin, without any sharp boundary, although maximum concentration would remain at the origin. The average distance from the origin would increase continually, but at an ever-decreasing rate, being proportional to the square root of the number of days involved. Thus, in 4 days the magnetic points would be scattered, on average, over a radius of $2 \times 15 = 30$ Mm; in 100 days, the radius would be $10 \times 15 = 150$ Mm.

Two other types of motion also operate. One of these is differential rotation. Without this, the distribution would be circularly symmetrical (Fig. 43). In reality, differential rotation shears the field as shown in Fig. 43 (*b*). Even so, as much flux would move southwards as northwards. While this is so for a short while, it does not happen for long; the flux drifts away from the equator towards the pole. The reason is a polewards flow of gas in each hemisphere, recently observed by the American Thomas Duvall, using delicate studies of Doppler shifts. Between 10° and 50° latitude, the speeds are nearly constant at about 20 m/s. Other techniques have shown the flow extending to about 70°. No such observations have been possible outside these latitudes, but by symmetry the flow must drop to zero at both equator and poles. In the main flow, magnetic elements are carried bodily

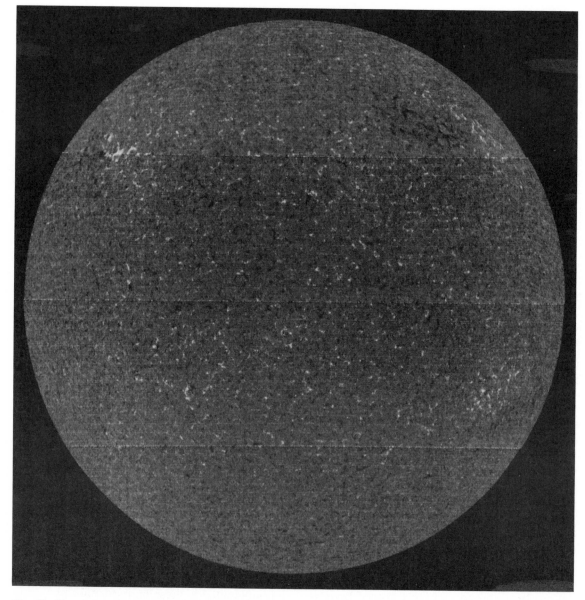

Fig. 41. *Magnetogram near sunspot minimum (28 February 1977). The field is concentrated into tiny magnetic elements. The distribution of polarities is random except near the poles, which are effectively unipolar, negative at the north pole and positive at the south. (The horizontal streaks are artefacts.)*

Fig. 42. *Magnetogram near sunspot maximum (23 June 1980). Strong fields occur in active regions, and large parts of the surface are occupied by unipolar regions having a characteristic shape, aligned polewards and eastwards in both hemispheres. Within these, the fields often outline clearly the boundaries of the individual supergranules.*

47

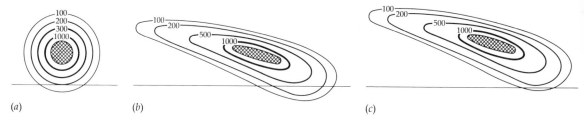

Fig. 43. *Schematic diagram of the transport of flux across the Sun's surface (northern hemisphere). Numbers indicate contours of equal flux density (arbitrary units). (a) Flux distribution after 0.3 year with transport by a random walk alone (daily breakdown of supergranules). The inner, hatched circle contains one-half of the total flux.*
(b) Differential rotation added to the random walk. Note the way that the flux is drawn out in a direction slanting polewards and eastwards.
(c) Polar flow added to the random walk and differential rotation. The pattern as a whole migrates polewards. Close enough to the equator, differential rotation and poleward flow become very small and transport is dominated by the random walk.

polewards about 13° in 3 months, in which time the average radius of the random walk becomes about 12°. Up to about this time, the field can expand by random walk towards the equator. Subsequently the steady poleward flow dominates as the rate of random-walk expansion diminishes. Thus the flux is not only skewed by differential rotation but, on average, moved towards the poles.

So far we have considered only what happens if flux is liberated from a single sunspot; but spots occur in bipolar groups, so that flux of opposite polarities is liberated from leader and follower. While these are separated in longitude, the leader is also about 2° closer to the equator on average. We can trace the development of skew unipolar flux regions from each of these spots separately. As we might expect, they overlap. Observation shows that, despite at times a narrow zone of mixed polarities between them, the Sun is able to cancel opposite polarities over a reasonable length of time – otherwise the unipolar regions would not be so well defined (we shall see in Chapter 7 how it does this). The net result is that the processes described (the random walk, differential rotation, the poleward flow and the gradual cancellation of opposite polarities) cause the development of the typical unipolar fields found around sunspot maximum. A rather nice way of showing this has been developed at Kitt Peak National

Observatory by John Harvey, who has prepared magnetograms showing the average field distribution over the *whole* Sun for each rotation (Fig. 44).

With more and more flux liberated from sunspots, the amount on the Sun's surface increases by a factor of about four times from sunspot minimum to maximum, if we ignore the sign of the flux. This is not nearly as great as if there were no cancellation of flux where positive and negative fields overlap.

Where the packing of magnetic elements is dense, we find an elegant network structure. While we have just described this as a concentration near the boundaries of the supergranules, some 30 Mm in diameter, a closely packed network always has a finer structure superimposed. This is due to the existence of convection on scales intermediate between supergranules and granules (see for example Fig. 21). Although average velocities in these smaller cells are not as great as in supergranules, there are regions near the supergranule boundaries where the motions of the various sizes of cells balance. Here the magnetic elements can be trapped, outlining the smaller structures.

The magnetic elements themselves are smaller than have been resolved by magnetographs to date. Many different types of observation agree in showing that the field strengths lie in the range 1000 to 2000 gauss. Fields of such strength, at least half of those

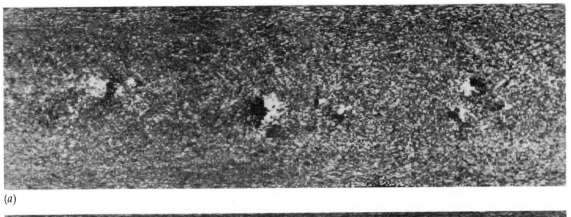

(a)

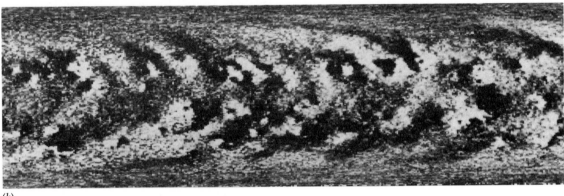

(b)

Fig. 44. Average field distribution over the whole Sun during a rotation (27 days). The Sun is shown on a rectangular projection such that longitude is on a uniform scale. Latitude is distorted so that equal areas anywhere on the Sun have equal areas on the diagram. (a) Near sunspot minimum (July 1975); (b) near sunspot maximum (October 1980). Notice in particular the much greater number of active regions near sunspot maximum, and the presence of unipolar regions of both polarities. These are virtually non-existent near sunspot minimum, when the fields are almost invariably of mixed polarity (see Fig. 41).

found in the strongest sunspots, occur in little tubes whose diameters seem to be perhaps only 100 to 300 km. Overall, they occupy a very tiny fraction of the Sun's surface. Yet at sunspot minimum, and to an appreciable extent at sunspot maximum, the magnetic field in the higher corona and in interplanetary space comes from these elements (the sunspots themselves also contribute to this field). As it is difficult for gas to pass across lines of magnetic force, all the gas in the corona and in the solar wind (which escapes from the Sun into interplanetary space along lines of force) comes either from the magnetic elements or from sunspots. Therefore the properties of the unresolved elements have become important for they decide, to a considerable

extent, the properties of the upper solar atmosphere, interplanetary space, and the nature of solar-terrestrial relations.

Special techniques, based on the way that magnetic fields show Zeeman splitting of spectrum lines (see Plate VIII), have been developed to look for motions *inside* magnetic elements, even though the elements cannot be resolved in the best telescopes. Two forms are found: (1) a systematic downflow of gas within the tubes of force; and (2) oscillations rather similar in period (e.g. 5 minutes) and amplitude to those in the non-magnetic atmosphere (see Chapter 4). The systematic flow increases downwards from 0.1 km/s or less in the chromosphere to nearly 1.5 km/s low in the photosphere. It is present, it seems, in *all* magnetic elements, even in both

49

components of very close bipoles known to be connected by small looped tubes of magnetic force. This should drain all gas from the loop in a few minutes, but the downflow may continue for a day or more. The Sun obviously has the ability to replenish gas inside the loops, even though it should be impossible for highly ionized plasma to cross from the non-magnetic exterior into the tube. The answer seems to be that these tubes pass through a cool region in the atmosphere where there is a considerable pool of neutral gas; this can diffuse across field lines into the low gas pressure of the tube.

As to the oscillations, the amplitude increases upwards. Oscillations of the field above magnetic elements will turn out to be related to the heating of the upper atmosphere, to be described in Chapters 8 and 9, and we shall be considering at that stage what the various options are.

We have spoken of the way that there must be a balance between the external gas pressure and the combination of gas pressure within the magnetic field and the magnetic pressure. Near the surface, where the gas pressure drops off very rapidly with height, the magnetic pressure must drop off rapidly also. There is only one way for this to happen: by the tube diameter expanding with height. If the tubes are isolated, the expansion with height is symmetrical about the tube axis, but if the tubes are packed fairly closely, there is room for expansion only at right angles to the network axis, i.e. to the supergranule boundary. In this case the field spreads out much more rapidly with height.

We can study this process by looking at magnetic fields well away from the centre of the Sun, so that the field lines may have strong components along the line of sight; and by using spectral lines that originate high enough in the atmosphere. Now a surprising result appears. Plates IX and X show three-dimensional representations of the field in two active regions. As expected, it is highly localized low in the atmosphere, but higher it expands diffusely towards opposite polarities some distance away, forming magnetic canopies presumably with fairly sharp bases, below which there is no magnetic field. This phenomenon is not restricted to the immediate neighbourhood of sunspots – it occurs anywhere that there is an appreciable concentration of network field. Typical heights of canopy bases are 500 km or so; although they are variable, the bases are almost horizontal over large areas.

The significance of magnetic canopies is that the need for pressure balance requires a big drop in gas pressure on passing upwards through the base of the canopy. This is rather an embarrassment at present, for all studies of the temperature and pressure variations with height in the Sun's atmosphere have taken no account of the possibility of sudden drops in gas pressure at the 500 km level. In consequence, the results must be very dubious. In particular, there is usually said to be a temperature minimum around 550 km, but the nature of this minimum is now highly suspect, and there is urgent need to analyse the atmospheric structure taking magnetic effects fully into account.

7 The sunspot cycle – *what and why*

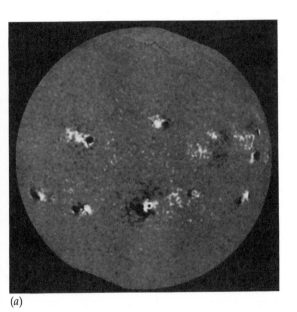 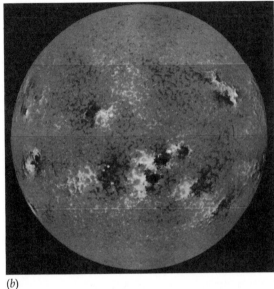

(a)   (b)

Fig. 45. *Magnetograms from successive cycles demonstrating Hale's laws. (a) All the spots in a given hemisphere, north or south, have the same east–west alignment of polarities within the group during the whole 11-year cycle. The alignment of polarities is opposite in opposite hemispheres. (b) The alignment of polarities reverses in successive cycles.*

The spottedness of the Sun is variable. On average, it rises to a maximum in about 4 years and drops to a minimum in about 7 years, providing a cycle roughly 11 years in length. Despite detailed fluctuations on a daily, monthly or even yearly scale, there is remarkable order in the major aspects of the cycle.

The first of these is the magnetic dominance of everything that occurs. This is expressed in Hale's laws (see Fig. 45):

1. Throughout the whole of a cycle, from sunspot minimum to the next sunspot minimum, all the spot groups in, say, the northern hemisphere have the same magnetic alignment, i.e. all leader spots have identical polarities, follower spots have the reverse polarity. In the southern hemisphere, the sense of polarity is the opposite.

2. In the next cycle, the magnetic polarities reverse in each hemisphere.

There is therefore a magnetic cycle 22 years long, consisting of two 11-year sunspot cycles.

A second regular feature is the variation in average sunspot latitude with the cycle. The first spots of the new cycle lie around 30°N or 30°S latitude, on rare occasions as high as 40°. As the cycle progresses, the average sunspot latitude decreases, and by the end of the cycle it is more like 8°N or 8°S. A plot of the latitudes of spots with time is known as the Maunder butterfly diagram (Fig. 46), so called after the discoverer of the

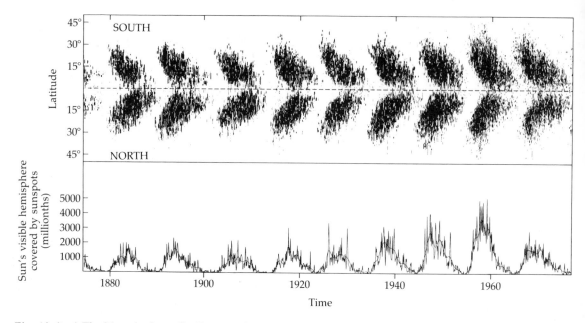

Fig. 46. (top) *The Maunder butterfly diagram, showing the latitude range of sunspots as a function of time. The first spots of a new cycle are at about 30°N or 30°S latitude, the last break out almost at the equator. (below) Millionths of the Sun's visible hemisphere covered by sunspots as a function of time. Note that cycles have been stronger in recent times than 100 years ago.*

phenomenon and from its appearance.

The first spots of a new cycle are distinguishable easily from the last of the old by two differences – polarity and latitude. It is not uncommon for spots of the new and old cycles to be present simultaneously for a year or two near sunspot minimum.

A third important property of the cycle is the reversal of the polar fields around the time of sunspot maximum. Magnetic fields are present not only in sunspots but are distributed widely over the surface of the Sun. Their properties were discussed in detail in the previous chapter, but an aspect of interest to the cycle is that on the whole, but not necessarily at any one time, the north and south poles of the Sun have opposite magnetic polarities. The polarities reverse around sunspot maximum, but by no means in a definite way, and they may oscillate several times before settling down. In consequence, both north and south poles may have the same magnetic polarity at times. However, on a broad scale, the phenomenon of polarity reversal is well established.

A fourth interesting property is the torsional oscillation of the Sun, or oscillation in its rate of rotation, discovered recently by Robert Howard and Barry LaBonte at Mt Wilson Observatory in California (Fig. 17). At any latitude the rate of rotation oscillates with an 11-year period, sometimes faster than average, sometimes slower. The time of the maximum rate of rotation varies with latitude. An oscillation starts at about the same time in the two polar zones and drifts down to the equator in 22 years. New-cycle sunspots break out soon after the maximum rate of rotation arrives at about 30°N or 30°S. Clearly the rotational oscillation is connected closely with both the 11-year sunspot cycle and the 22-year magnetic cycle.

The history of the sunspot cycle can be traced directly back to the time of the discovery of sunspots, but with decreasing accuracy for observations before about 1850 and even more so from 1610 to 1700. The curve shows clearly that the strength of the cycle is quite variable. Many astronomers have attempted to explain these variations by supposing that there are other longer periods

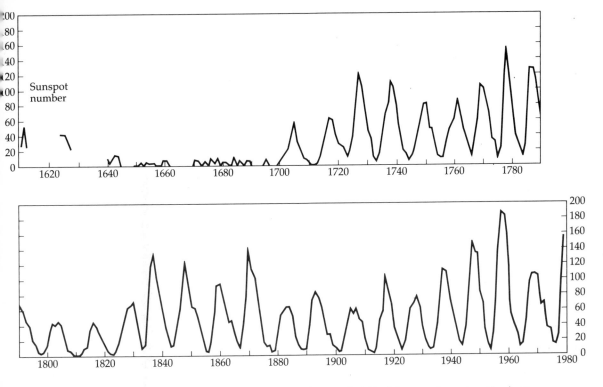

Fig. 47. Annual mean sunspot numbers from 1610 to 1979. The so-called Zurich sunspot number, R_z, is an artificial concept, assessed by the total number of sunspots visible in a standard type of telescope plus ten times the number of separate spot groups.

of smaller amplitude superimposed on the 11-year cycle. The most popular has been of about 80 years, called the Gleissberg cycle, although it was discovered not by Gleissberg but by Wolf. The reality of these has been controversial. However, analysis by George Williams of 1760 years of Late Precambrian annual deposits (varves) in South Australia (see Chapter 12), laid down about 680 million years ago, has shown that the weather in that region was sensitive to the sunspot and magnetic cycles. The time span studied is many times greater than that for which direct solar measurements have been made, yet the 11- and 22-year cycles were the same then as today, showing that they are permanent features of the Sun. A striking feature was the extensive sequences of alternating strong and weak sunspot cycles (see Chapter 12). Longer-period variations in activity also occurred, with periods of 145 and 290 years, and a weaker cycle of period around 90 years.

There can be no controversy, either, about the lengthy interval, shortly after the discovery of sunspots, when from about 1640 to 1700 there were very few spots on the Sun. Known as the Maunder minimum, after the English astronomer who first drew attention to its unusual behaviour, this virtual absence of spots occurred at about the same time as a long spell of cold weather in Europe, the 'Little Ice Age'. The effect of sunspot activity on the weather is discussed in greater detail in Chapter 12.

How can we account for the cycle? All attempts have been extremely controversial. Yet we now have sufficient clues from the many various phenomena observed at the surface to form a consistent and satisfying explanation.

Sunspot formation is a cyclic phenomenon, and we may start our

53

Fig. 48. *Whenever flux strands of opposite polarities are dragged or pulled into contact, they reconnect rapidly to form a 'U'-shaped upper strand, which eventually floats out of the Sun, and a lower loop which is completely submerged. The ends of this loop are connected back to the original flux rope or ropes from which the sunspots were formed.*

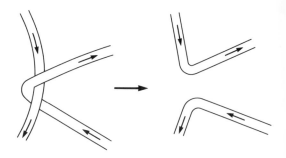

explanation at any convenient phase, for example sunspot minimum when the Sun's magnetic field has the typical appearance of Fig. 41. As at all times, the field is confined to tiny elements some 100 to 300 km in diameter and strength 1000 to 2000 gauss. They are distributed almost at random near sunspot minimum; the polarities are also random, except at high north and south latitudes which are magnetically unipolar.

New-cycle sunspots appear first at latitudes around 30°, the average latitude then drifting towards the equator over about 11 years. When the sunspots decay, after lives of typically a week to 10 days, each group releases magnetic flux of both polarities in equal quantities, as described in Chapter 6. This flux also takes the form of tiny magnetic elements, which are transported by gas motions to lodge near supergranule boundaries where horizontal velocities are small. When the supergranule breaks down after about a day, the magnetic elements are carried by gas flow to the boundaries of new supergranules, a process which is repeated endlessly. The individual flux strands diffuse in all directions across the Sun's surface by this process.

Inevitably, strands of opposite polarities will be dragged into contact (Fig. 48). We have already stressed that magnetic fields are effectively unable to move through the electrically highly conducting solar gases, but when the separation becomes small enough between opposing flux tubes aligned in opposite directions, or even between flux tubes crossing at appreciable angles, the tubes can reconnect in a few hours. In this way, two flux strands of opposite polarities which are dragged together will reconnect to

form an upper U-shaped strand, both ends of which pass out through the surface, and a lower loop completely submerged, its ends being connected back to the flux ropes from which the sunspots were formed. Eventually the U-shaped tube floats out of the Sun. This continuing process reduces greatly the flux of each polarity present on the Sun. It also leaves large areas of predominantly leading or following polarity, the 'unipolar' regions.

Not long after the breakdown of a spot group, the poleward flow of some 20 m/s starts to predominate, carrying what is left of the flux to polar regions in about a year. While this is happening, the flux strands at any one latitude rotate around the Sun at a rate appropriate to that latitude. Since the rate of rotation decreases from the equator towards the poles, the predominantly unipolar areas lag further behind at ever higher latitudes, producing the skew-shaped regions of Figs 42 and 44*b*. Fig. 49*a* shows, at successive times, their appearances to an observer rotating with the Sun.

The poleward flow is also responsible for changing the fields in polar regions. In either hemisphere the leader spot of a group has the same magnetic polarity as the polar field had at the beginning of the cycle; but on average the leader spot is about 2° closer to the equator than is the follower. Thus, leader flux reaching polar zones lags behind follower flux by the time taken to drift 2° at the poleward rate of 20 m/s, about 14 days. In addition, flux is released from the follower spot rather quicker than from the leader, increasing the lag by another few days. In the rising phase of the cycle, when the rate of flux release is increasing with time, the follower flux arriving at a given high latitude

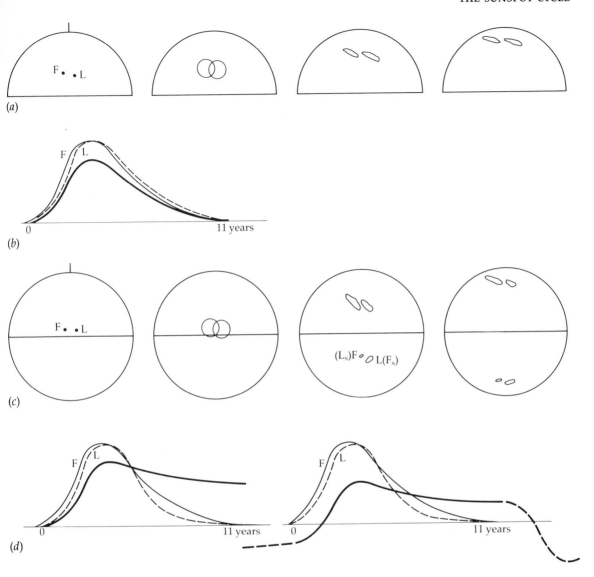

Fig. 49. *Reversal of magnetic polarity in polar regions. (a) (left to right) Magnetic fluxes released from leader L and follower F spots diffuse outwards, cancelling where they overlap to leave mainly unipolar regions which drift polewards. The follower spot starts at a slightly higher latitude than the leader, and follower flux arrives in polar regions first by two weeks or so. (b) Rate of arrival of follower F and leader L flux in polar regions (vertical axis) during the cycle. The two curves are identical (but displaced by two weeks or so) unless there is loss of flux of either polarity. More F flux than L arrives up to sunspot maximum, but the reverse happens later. The heavy curve shows (to a different vertical scale) the accumulated flux vanishing at the end of the cycle. (c) Later in the cycle when spots appear nearer the equator, some flux diffuses from, say, the northern to the southern hemisphere, more L than F because the leader is closer to the equator. Poleward drifts carry the remnant unipolar regions towards the respective poles, less L flux than F going northwards but more L than F going southwards where the polarities are the same as southern follower (F_s) and leader (L_s). Diffusion of flux from southern to northern hemisphere doubles flux accumulation near the north pole during the cycle. (d) (left) Rates of arrival of F and L fluxes in polar regions during the cycle, allowing for loss of flux across the equator. The heavy curve shows (to a different vertical scale) the accumulation of flux, which is doubled by flux diffusing across from the opposite hemisphere. (right) The polar field positioned so that its magnitudes are equal but of opposite sign at the ends of successive cycles.*

Fig. 50. Flux tubes drawn out by gas circulating in the convection zone between equator and polar regions. Because the density is greater in the deeper regions, the equatorward velocity is only about one-tenth of the poleward velocity. It takes about a year to flow from sunspot zones to polar regions, and about 10 years from there to the higher sunspot latitudes or 20 years to near the equator.

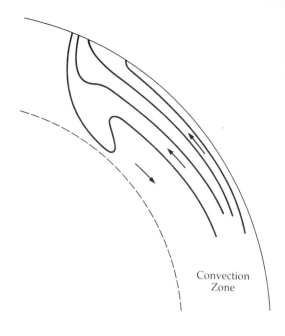

Convection Zone

corresponds to a slightly later and stronger phase than does the leader flux. Therefore on average it exceeds the leader flux. The difference may only be small, but it is sufficient eventually to reverse the magnetic polarity near the poles.

Fig. 49b shows how, after sunspot maximum, we might have expected the same lag to cancel exactly the flux carried to polar regions. However, many spots are then close to the equator, at latitudes of the order of 5° or so, where the poleward flow is slower. Some of the flux can diffuse across the equator, more from leader spots than followers because the latter are further from it and their flux is subjected to a stronger poleward flow. Once across the equator, poleward flow in the other hemisphere carries that part of the flux towards the opposite pole (Fig. 49c). Thus in the declining phase of the cycle the excess of leader flux carried towards its own polar region is not as great as otherwise expected, and follower flux may even continue to accumulate there (Fig. 49d (left)). At the end of the cycle, the polar field remains reversed from its initial polarity (Fig. 49d (right)).

An inevitable consequence of the poleward flow in the outer solar layers is a deeper return flow; otherwise the Sun would become grossly distorted in a very short time. The circulation is almost certainly confined to the convection zone, within 200 000 km of the surface, as shown in Fig. 50. The counterflow or equatorward velocity, which can be calculated only roughly, is much slower, some 1 to 3 m/s or about one-tenth of the surface poleward velocity, so that it takes about 10 years to flow back to the higher sunspot latitudes around 30°.

The flux strands remain projecting through the surface in polar regions unless they happen to reconnect with strands of opposite polarity. In any case, the submerged parts of the strands are carried down by the circulating gas. During its slow passage back towards the equator, part of a strand at lower latitude rotates more rapidly around the Sun than a part at higher latitude because of differential rotation. Thus every strand tilts further and further from the north–south direction as time goes on and as it moves equatorwards. This brings the strands closer and closer together.

The concentration of flux strands in this way is accompanied by the formation of flux ropes. Turbulent motions in the gas drag neighbouring strands into frequent contact, tending to produce tangles as in a beginner's fishing line. But the strands reconnect rapidly, producing flux strands which are more nearly parallel and close together. There is one exception to the reconnection process. If tubes cross only at an exceedingly small angle, say a fraction of a degree, the reconnection time becomes exceedingly long. Thus very gentle twists survive, and large numbers of flux strands can be twisted together very gently to form a rope of flux

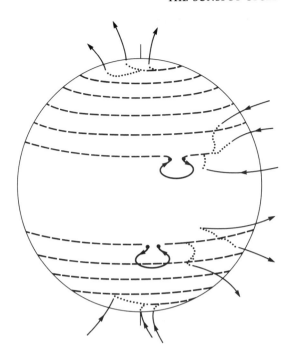

Fig. 51. Directions of the subsurface field (broken lines) and polarities of sunspot groups compared with polar fields earlier in the cycle. External field lines are shown as unbroken curves. Field lines disperse at lower latitudes, and ultimately pass out through the surface where they appear as tiny magnetic elements. Loops show where flux has emerged from sunspot groups.

strands, just as a thread of textile is formed from fibres by twisting. Because of the randomness of turbulence in the convection zone, the number of fibres in a rope will vary along its length; but the average number should increase as migration towards the equator increases the concentration of flux strands.

Not all the flux strands initially migrating equatorwards are of the same polarity. Many are loops remaining after the earlier reconnection of flux strands of opposite polarities, but these loops are gradually annihilated. Ultimately, there remain only the strands connected to the polar field.

At some stage a dramatic change occurs. Because part of the pressure inside a magnetic tube is magnetic, there is less gas there than outside. The tube tends to float. Whether it succeeds depends on its size. The upthrust on a given length is proportional to the cross-sectional area, i.e. to the square of the radius. The gas resistance is proportional only to the radius. So the bigger the tube, the easier it is to float upwards. Single strands are too small to do this, but when a particular part of a rope has enough strands it may float upwards, reaching the surface after a few months. When it breaks the surface, a pair of sunspots forms as described in Chapter 5.

We know too little about flux-rope formation to predict the length of time needed for floating after leaving polar zones. It is certainly a chancy business, depending on randomness in rope formation. It also depends on the initial concentration of flux in the polar regions, which varies with time, taking longer on average if the initial strand concentration is weaker. Hence we expect

spots to break out with minimum delay, i.e. at highest latitudes, when they originate from the transport and winding of maximum initial polar flux. As time goes on, the average latitude of sunspot formation drifts down to very low latitudes but at a slowing rate, since the circulation velocity slows to zero at the equator.

Observation has shown that new-cycle sunspots break out at about latitude 30° some 8 years after reversal of the polar field, and the average sunspot latitude then decreases until the process is exhausted after about another 11 years. This is in good agreement with what we would predict, so good that we can be confident of the explanations given and use the observations to improve our knowledge of the deep circulation velocities. When this is done we can reproduce the variation in sunspot latitude through the cycle very accurately.

What about the alignment and polarities of the sunspots? On floating upwards, the flux rope is strong enough to maintain its original orientation rather closely. As seen from Fig. 51, the spot group is aligned almost east–west, the leader spot being slightly closer to the equator. The leader has the same magnetic polarity as the polar field from which it was generated, i.e. the polar field after the previous sunspot maximum and at the end of the previous cycle. In successive cycles the sense of polarity in spot groups will reverse. In opposite hemispheres the polarities of the spot groups will also be opposite, since the polar fields are of opposite polarities. Thus Hale's polarity laws (see page 51) are explained fully. Clearly, the magnetic cycle is equal to two ordinary sunspot cycles, approximately 22 years. Its duration is decided by that of the circulation, but this is not yet understood quantitatively.

An aspect of the cycle still to be explained is the torsional oscillation described earlier. Differential rotation winds the flux strands around the Sun. The wound strands react so as to reduce the differential rotation slightly. The effect of this is greatest where the concentration of wound strands is greatest. There will be negligible effects near the equator, where differential rotation is negligible. It can be shown that, at a fixed latitude, the increase in rotation rate is greatest some 1.5 to 3 years before the wound field is strongest, i.e. about 1.5 to 3 years before maximum sunspot activity at that latitude. Finally, the latitude where the change in rate of rotation is greatest drifts down at the rate of the deep, equatorward return flow. The surface rotational changes follow closely the changes in rotational rates deep in the convection zone because vertical convection spreads such changes to the surface rapidly.

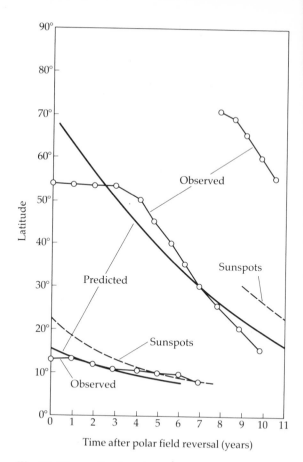

Fig. 52. *The predicted variation in rotation rate at different latitudes during a sunspot cycle, compared with those observed by Howard and LaBonte. The predicted delay between the maximum rate of rotation and the outbreak of sunspots, shown by the broken lines, also agrees well with that observed.*

The patterns of predicted and observed changes in rotation rate are compared in Fig. 52. The agreement is remarkably good from the equator up to about latitude 55°. At higher latitudes, differences appear between prediction and observation. It is unfortunate that the time over which observations have been available is so short; we cannot decide yet whether residual differences are due to systematic variations in flow with latitude, which would provide information on the deep circulation in polar regions, or whether they are just random.

8 Coming out from the Sun
– *first the chromosphere*

'Chromosphere' means 'coloured zone'. It appears for a few seconds as a magenta fringe at the beginning and end of a total eclipse of the Sun. Lying immediately above the photosphere, it is too rarefied to emit white light, but it can still emit spectral lines. The density falls off rapidly with height, e.g. by a factor of 100 in the first 500 km. Most of the visible spectral lines disappear within 1000 km, leaving only a few of the very strongest lines, e.g. the red Hα and blue Hβ lines of hydrogen and the violet H and K lines of calcium. These give the chromosphere its colour.

The most remarkable property of the chromosphere is the complete dominance of its structure by the magnetic field, resulting in features of great beauty entirely different from any encountered at lower levels. This is due to the very low gas pressure there. Even the limb chromosphere, whose thickness is highly irregular, from perhaps 1000 to 7000–8000 km, shows spicules (Fig. 53) which owe their shapes to the magnetic field. Usually the spicules lie within about 45° from the vertical.

The chromosphere can be observed readily on the Sun's disk with a spectroheliograph or a narrow-band filter, both of which isolate light from a spectral line and produce images of the chromosphere without the Sun's blinding glare. Of all lines in the visible spectrum, hydrogen Hα is used most often. Figure 54 is a spectroheliogram obtained with this line in 1967, early in the sunspot cycle, so that we can identify features typical of both the active and the quiet Sun. In it are three quite different types of structure – long dark filaments, bright plages, and a jungle of almost horizontal fibrils covering the whole surface. The filaments are not part of the chromosphere, but are prominences seen on the disk – huge volumes of cool, denser hydrogen gas extending out to 50 000 km or more from the surface. Their locations, shapes and very existence are the direct result of the magnetic field. We discuss them further in Chapter 10. The bright plages are located mostly near sunspots. These are the active regions where brilliant, short-lived flares occur from time to time. Dotted over the surface are similar but smaller, fainter structures known as plagettes, which have migrated from former active regions. Plages and plagettes coincide with strong concentrations of magnetic field.

Most of the chromosphere is fairly quiet. The fibrils are very obvious near plagettes, extending outwards on one or both sides (Fig. 55). Near the plagette axes they are usually bright at the Hα line centre, indicating higher than average temperatures, but they become darker further away. Velocitygrams (Fig. 56) show that the gases are in rapid motion, the numbers of fibrils exhibiting inwards and outwards motions being about equal. If this seems to conflict with the velocities found in magnetic elements, we shall see shortly that it is not so.

Deeper regions can be seen when the filter wavelength is moved into the wings of the absorption line (Fig. 57a, b). Fibrils still appear, now dark against the much brighter background, at 7/8 Å from the Hα line centre in either wing (Fig. 57c, d). The individual fibrils are quite different in opposite wings, showing that there are strong Doppler shifts and therefore big velocity components along the line of sight. The motion is actually along

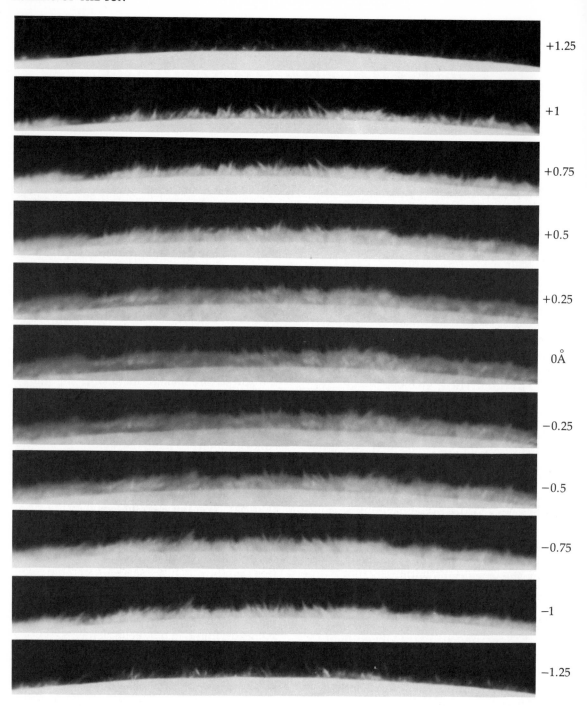

+1.25

+1

+0.75

+0.5

+0.25

0Å

−0.25

−0.5

−0.75

−1

−1.25

Fig. 53. *The chromosphere at the limb, at different intervals from the centre of the strong hydrogen Hα line. Near line centre (0 Å), the photospheric limb can still be seen in the above pictures, but this is an artefact due to the presence of a little stray white light; the chromosphere itself is quite opaque. The chromosphere is strongly spicular. The spicules in one wing of the line (say 0.5 Å or more from line centre) are quite different from those in the other wing. This is because gases move at high velocities along the spicules (up to 30 km/s or more), which are inclined towards or away from the observer, so producing Doppler shifts making a spicule visible in one wing only. The spicules lie along lines of magnetic force.*

Fig. 54. *The Sun in light from the centre of Hα. At this wavelength light can escape only from high in the chromosphere, although the level involved varies greatly from point to point. The long, dark filaments are prominences seen on the disk. Near spot groups are bright plages, where there are strong concentrations of magnetic flux tubes. Smaller brightish features (plagettes), also associated with strong fields, fleck the disk. The remainder of the surface is covered with long, narrow fibrils (particularly near the spot groups) and other finer structures.*

61

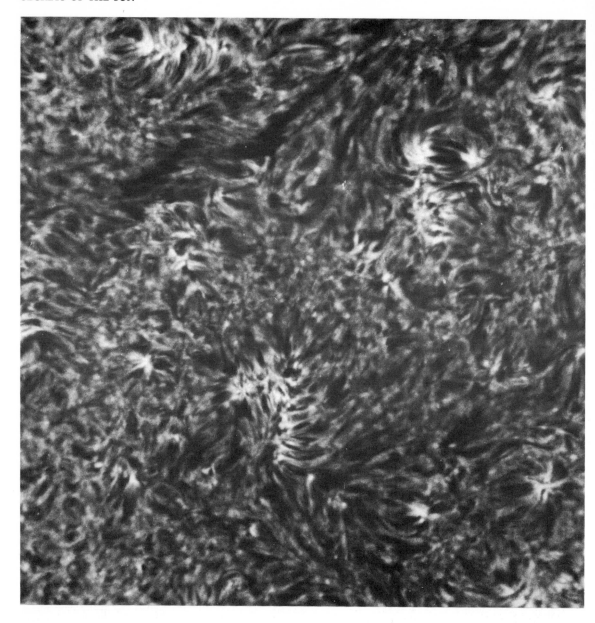

Fig. 55. *The Hα chromosphere in greater detail. This region near the centre of the disk contains some bright plagettes (wherever there are strong concentrations of magnetic elements along supergranule boundaries), from which fibrils extend on one or both sides. The roots of the fibrils are bright, indicating higher temperatures than elsewhere. The fibrils lie well above the fine chromospheric granulation seen at the centre of the filtergram; the latter structures seem to occur only in or over non-magnetic regions.*

Fig. 56. *(right) Velocitygrams of portion of the region of Fig. 75 obtained at 0.5 Å (a) and 0.75 Å (b) from line centre. Receding gases (descending) are displayed as dark, rising gases as bright. About equal numbers of fibrils carry rising and descending gases at any one instant. The direction of motion at a given point is subject to rapid change.*

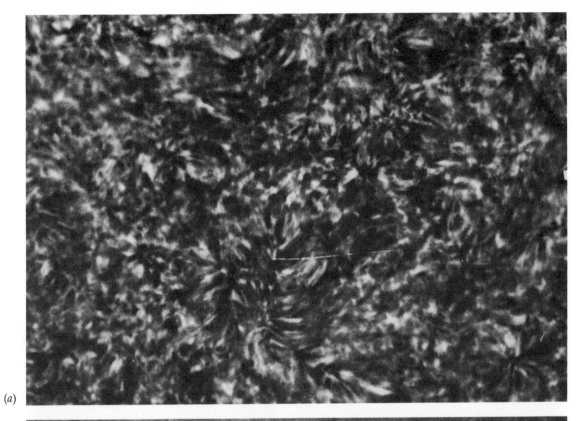

(a)

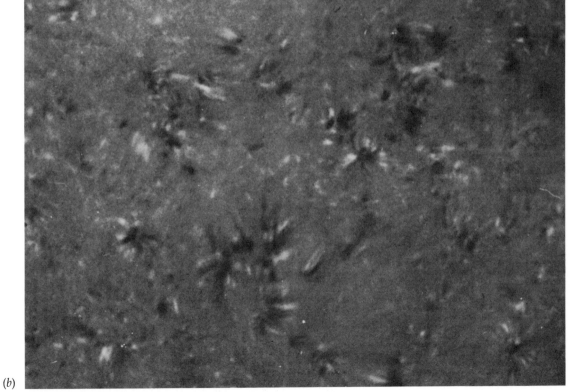

(b)

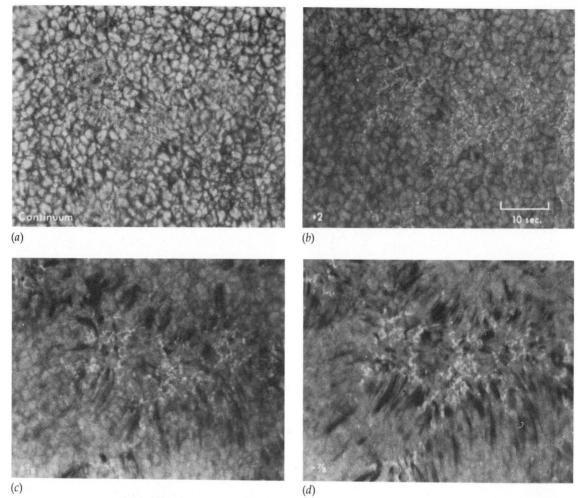

(a)

(b)

(c)

(d)

Fig. 57. (*a, b*) *The filigree is a pattern of bright features seen further into the wings of Hα. Clearest at 2 Å from line centre (b), it coincides with magnetic flux tubes. In white light (a), the granulation shows a curious loss of contrast in such places. The filigree is in a continuous state of change, individual features apparently jostling around with horizontal velocities of the order of 0.5 km/s.*
(*c, d*) *The deeper chromosphere as seen +7/8 or −7/8 Å from the Hα line centre.*

the almost horizontal fibrils, either towards or away from their footpoints. Further still from line centre the photospheric granules start to appear. The narrowest Hα structures, the filigree, are found about 2 Å from line centre and therefore deep in the atmosphere (Fig. 57b). These coincide precisely with the tiny magnetic elements; the whole Hα fibril structure is magnetic in origin.

Higher in the atmosphere, the gas pressure drops. As described earlier, the tubes of magnetic field then expand rapidly with height, reducing the field strength and magnetic pressure so as to maintain a

pressure balance. For this reason the dark fibrils in Fig 57c and d are much wider than the filigree of 57b.

We saw in Chapter 6 that magnetic canopies form typically at about 500 km or so above non-magnetic regions wherever the packing of magnetic elements along supergranule boundaries is fairly dense, e.g. in active regions and the stronger unipolar regions. Even where the elements are sparse, as in weak unipolar regions, canopies are present, though at somewhat greater heights averaging perhaps 700 to 800 km. The lower height in densely packed regions is due to

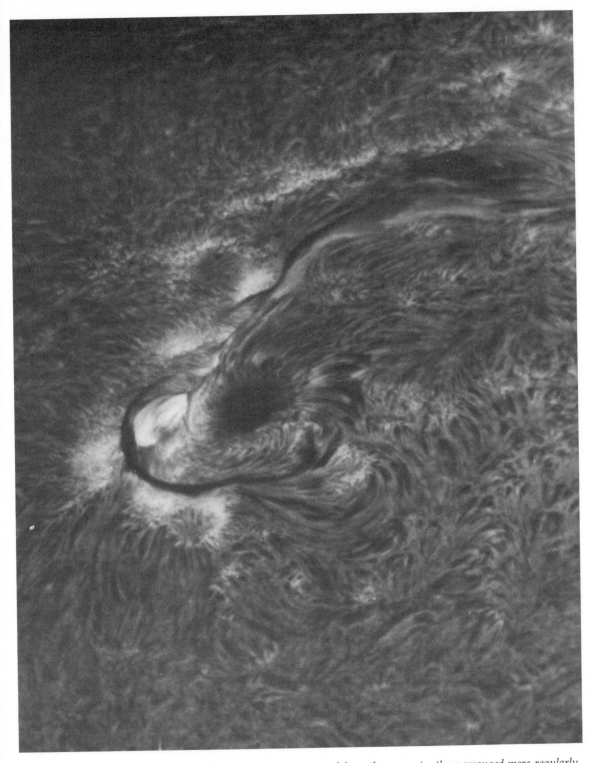

Fig. 58. *The chromosphere in an active region. Long fibrils extend from the sunspots, those arranged more regularly forming a superpenumbra, often around only part of the sunspot. Details of the bright plage structure can be seen, usually with shorter fibrils extending away from the spots. Almost all active regions generate prominences, seen as long, sinuous dark markings. Later these migrate towards the poles.*

Fig. 59. The Hα fibril structure around a simple sunspot near disk centre. The dark umbra is the same as seen in white light. The penumbra appears faintly, about twice the diameter of the umbra.

the interference of neighbouring elements with one another. Effectively, they can then expand only in one direction, at right angles to the supergranule boundary, and more rapidly with height. In the plage regions around sunspots the canopies are lower still.

The Hα fibrils, which lie above the canopy bases, give the chromosphere a peculiar fascination. Aligned with the magnetic fields, they are formed into long, sinuous curves near active regions (Figs 58 to 64). Their strange shapes resemble closely the field lines shown by iron filings near strong and weaker magnets (Fig. 60).

At the centre of Hα, absorption hides much of the structure. The haze disappears

on moving the filter transmission band away from line centre. Towards the limb, regions of strong magnetic packing then show an extraordinary array of structures, 4000 km or so high, projecting mainly outwards from the supergranule boundaries though with some at odd angles, their bases being pinned into magnetic elements (Fig. 65). At the limb these would appear as spicules, as in Fig. 53. Regions of this type are always unipolar.

The nature of the chromosphere has always been puzzling, but there are enough clues now for us to understand its origin. It is worth spending some time on this, since it is only from the magnetic regions that gases can escape to the interplanetary medium and the Earth. Here they produce various

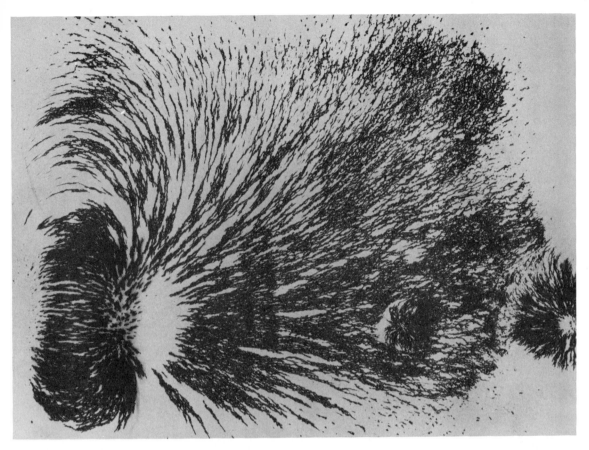

Fig. 60. Magnetic field lines near a strong magnetic pole (left) and two weaker poles of the same polarity (right), as outlined by iron filings. The lines of force near the weaker poles, in particular near the neutral points (just to the left of these poles), have the same patterns as many of the fibril structures in Figs 58, 59, 61, 62, 63 and 64.

solar-terrestrial phenomena which affect our lives. At present we still know too little of their causes.

Even solar astronomers often fail to appreciate the tremendous inhomogeneity of the chromosphere, and pretend that it is more or less uniform. It is rather like trying to describe a garden of tall trees, shrubs, flowers and lawn in terms of a layer of wood-chips.

The basic problem is to explain the properties of and differences between the tubes of magnetic force which are almost vertical, and appear as spicules, or almost horizontal and appear as fibrils. For the time being let us forget about motions and consider only what the magnetic fields would do in a static atmosphere. Later we can look

at the changes produced by motions. In this process we need to consider the relationships of the chromosphere to the photosphere below and to the corona above, particularly the narrow transition region between the chromosphere and corona.

The spicules and fibrils lie above the bases of the magnetic canopies – a good deal higher, judging from the photographs we have seen. They are parallel to lines of magnetic force because it is easy for gas to flow along lines of force but not across them. Thus the structures are aligned readily along the field. Very likely they originate in individual filigree features or individual magnetic elements – perhaps in only part of an element, but this is quite uncertain at present.

67

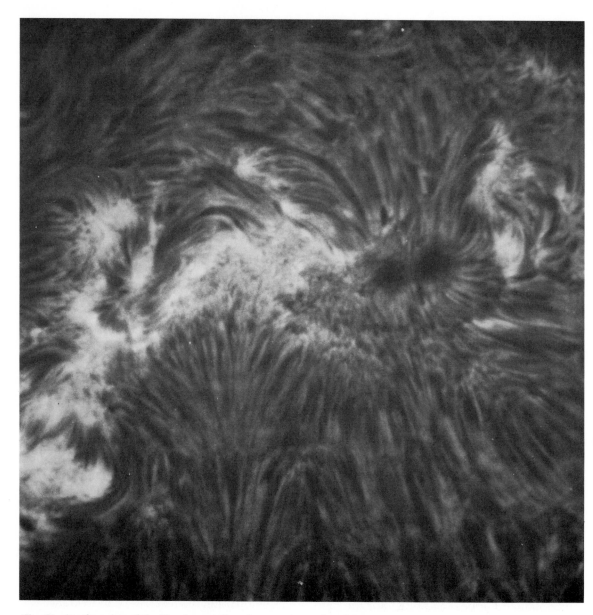

Fig. 61. *Another view of the Hα chromosphere in an active region. Many examples occur like those shown in the iron-filing patterns of Fig. 60 – these are due to the presence of clusters of magnetic elements in the photosphere below. Note a beautiful example of twisted flux strands just above the central bright plage.*

Plate XI. *'The Sun Tapestry' by Australian artist John Coburn. Flames surrounding the Sun on symbols of this type can be traced back to independent origins in America and Eurasia. Quite unlike the real Sun, they can only have been inspired by the corona during total eclipses of the Sun.*

Plate XII. *Petroglyphs from King's Canyon Wash, near Tucson, Arizona, USA. In the centre is a Sun-symbol consisting of a circle and a set of external radial lines. The latter almost certainly represent coronal streamers. Similar symbols are found in cave drawings in Spain and elsewhere.*

Plate XIII. *A modern Huichol (north-west Mexico) yarn painting showing the Sun. The radial spokes, found in many such paintings of the Sun, are similar to the ancient Red Indian petroglyphs to be found widely in North America. They are remarkably like the streamers of the outer corona.*

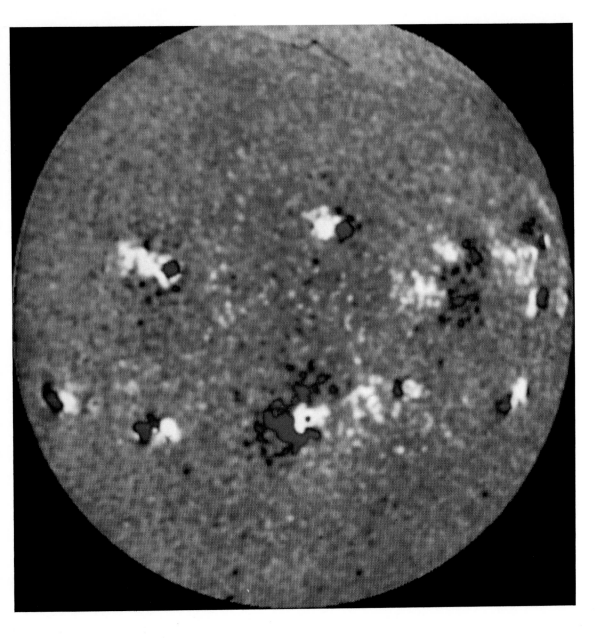

Plate XIV. *Magnetogram of 4 September 1973. Positive fields appear as white, negative as red.*

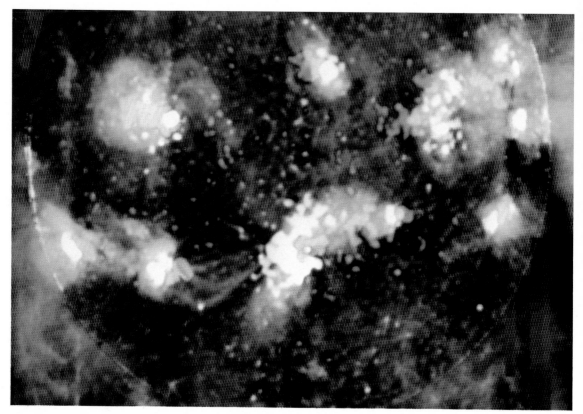

Plate XV. *An Fe XV image superimposed on the magnetogram of 4 September 1973. The positive fields now appear as white/pink, the negative as yellow/red. Fe XV emission appears as green or yellow-green. Long fibrils interconnect various active and other regions of strong photospheric field; these show the distributions of some of the field lines in the corona. The structure beyond the limb is confused by overlapping images.*

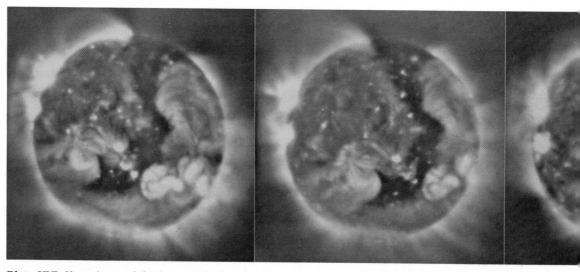

Plate XVI. *X-ray image of the Sun at daily intervals (the north–south axis is tilted slightly). Strong emission comes from over active regions. The dark region is a coronal hole, an extensive unipolar region from which field lines emerge into interplanetary space. It is much cooler than in the arcades of arches lining the boundaries of the coronal hole. Spread randomly over the disk are small, X-ray bright points, having lives of only about 8 hours.*

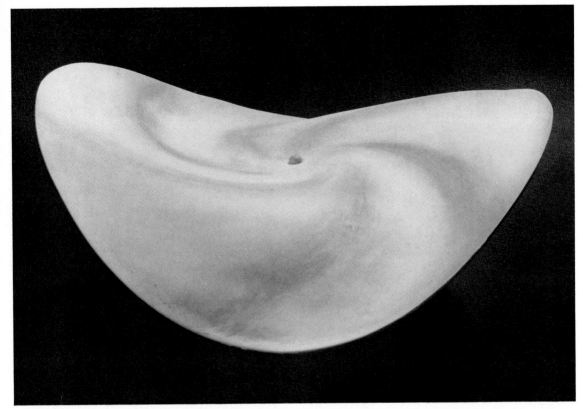

Plate XVII. Surface separating field lines spiralling outwards from and inwards towards the Sun, the yellow sphere at the centre. The Earth would be just inside the edge of the model. The general direction of the field lines is shown in blue and white. The surface is often known as the 'warped current sheet'.

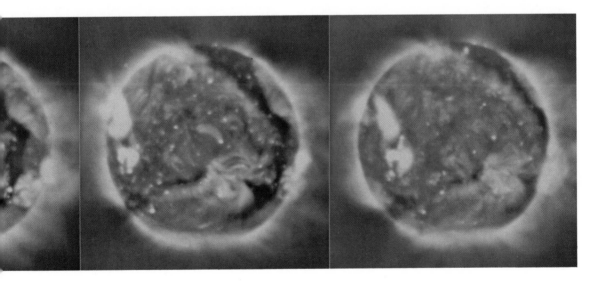

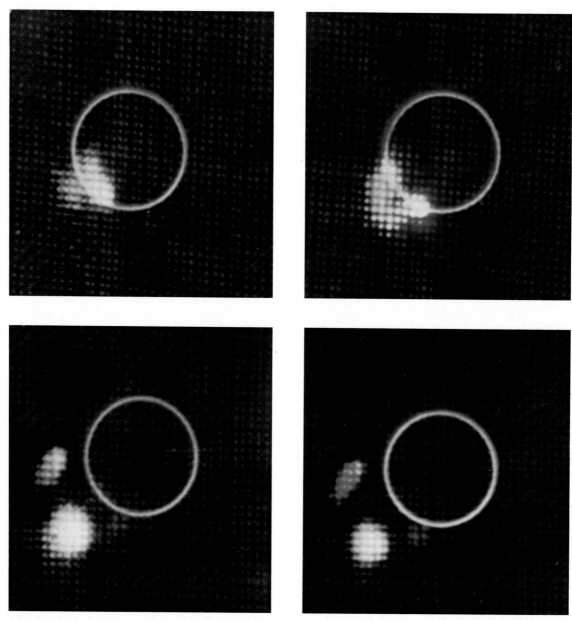

Plate XVIII. *A magnetic arch rising over a period of 50 minutes following a flare. Here the radio emission is from electrons gyrating around magnetic lines of force. The state of circular (or elliptical) polarization, shown as red or blue, depends on whether the field is inclined towards or away from the Earth. A white patch indicates that the field is perpendicular to the line of sight. This gives sufficient information for following the expansions of huge magnetic arches.*

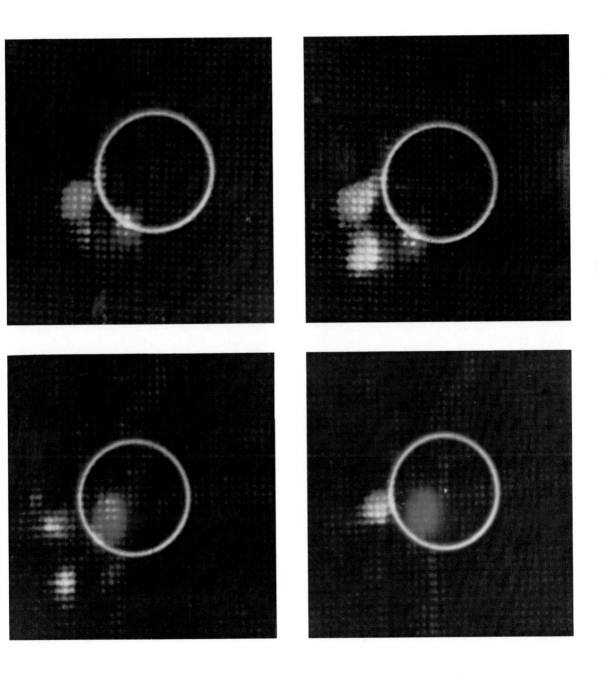

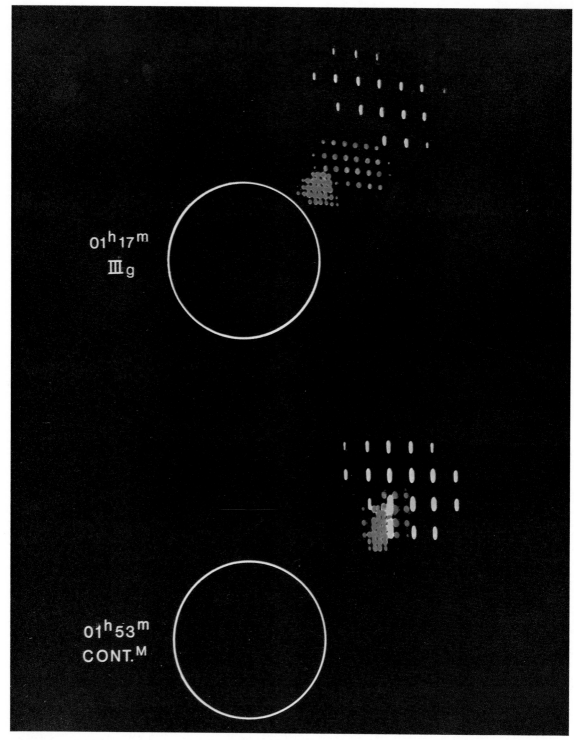

Plate XIX. (Top) Successive positions of a type III radio burst, observed at 160 MHz (red), 80 MHz (blue), and 43 MHz (yellow), on 22 August 1975. The heights correspond to the plasma frequencies, the velocity of ejection of the exciting electrons being about one-third the speed of light. (Bottom) Location of a moving type IV burst on 22 August 1975. Near its base, all frequencies are emitted together. In this case the radiation cannot be due to plasma oscillations but is rather synchrotron emission from weakly relativistic electrons.

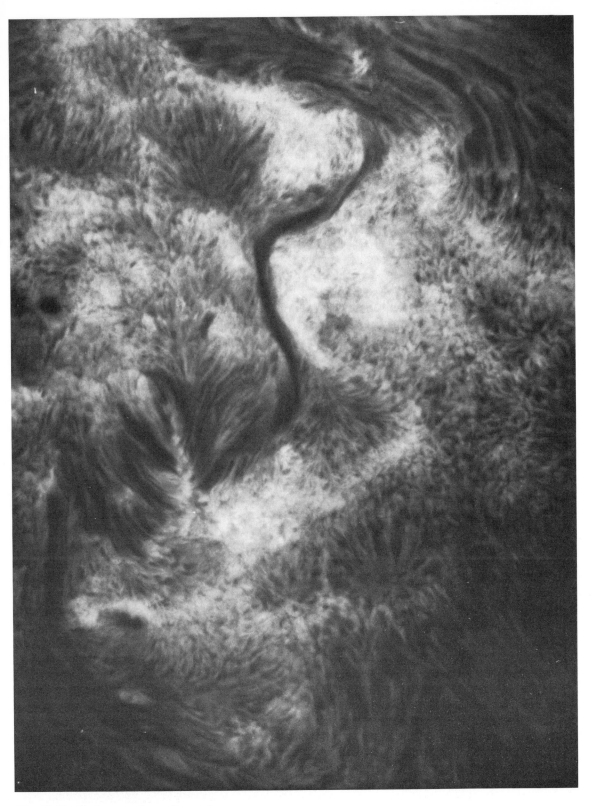

Fig. 62. Hα *fibrils around an active region.*

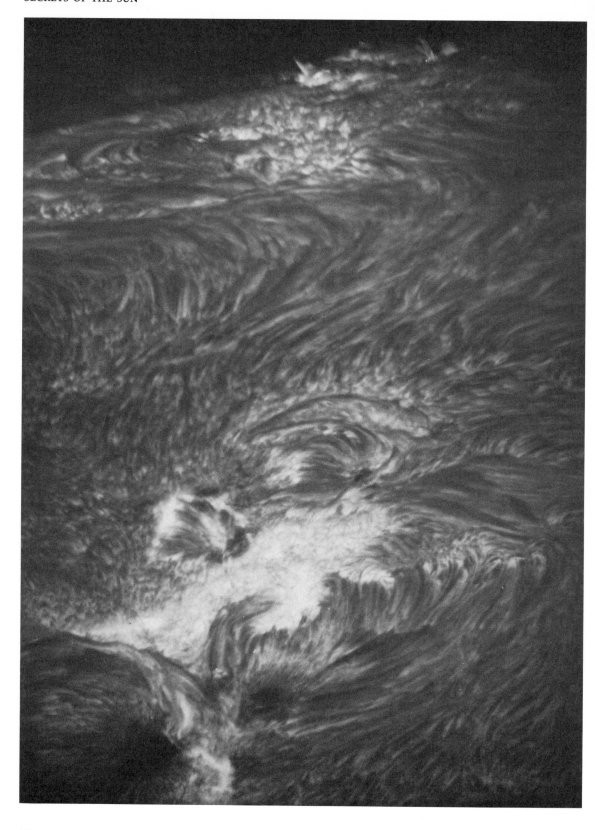

Fig. 63. (left) Hα *fibrils around an active region.*

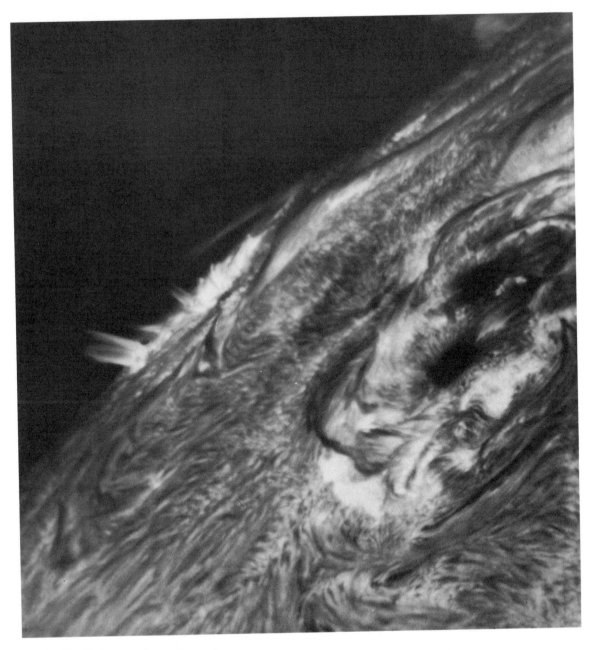

Fig. 64. Hα *fibrils around an active region.*

Fig. 65. *Fibrils observed 0.875 Å from the centre of Hα. These features would all appear as spicules at the limb (compare with Fig. 53 at 0.75 and 1 Å from line centre). The fibrils here outline supergranule boundaries. This photograph (13 February 1971) was obtained not long after sunspot maximum, and the region would have been unipolar. The fibrils extend 4000 to 5000 km upwards.*

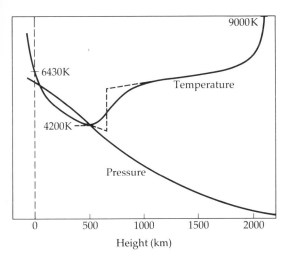

As mentioned in Chapter 1, it is not too hard to derive temperatures and pressures up to about 300 km in the photosphere from measurements of the brightness of the continuous spectrum. Things become more difficult higher up, because the chromosphere is seen only in light from atoms whose excitations there are understood much more poorly. Suitable observations are also difficult to obtain. Eclipse observations were used in earlier work. Observations from spacecraft have been far more popular in recent years; the advantage is the ability to study emissions in the far ultraviolet which originate at appropriate heights in the chromosphere and in the transition region between chromosphere and corona, but are absorbed in the Earth's atmosphere. The gas pressure has always been assumed hydrostatic, or nearly so – the pressure at every point just balancing the weight of the atmosphere above it. The net result has been a model of the average chromosphere such as in Fig. 66. The temperature drops to a smooth minimum of about 4200K near 500 km. Higher, it rises to some 9000K near 2000 km.

This sort of model has been available for a long time, and there has been an exasperating struggle to account for the outward rise in temperature. If heated solely by the Sun's radiation, the temperature should drop to a value not too different from the 4200K minimum without any subsequent increase. An increase requires some form of mechanical heating, for which the dissipation of acoustic waves has been invoked. Their amplitudes increase continuously as they travel higher into the less dense atmosphere. Eventually shock waves develop and these would dissipate the wave energy rapidly.

It is relevant that magnetic canopies occur in well-developed unipolar regions at heights of the order of 500 to 600 km, just where heating is first required by existing model atmospheres. Above the canopy base, the pressure is largely magnetic, so the gas pressure there is reduced greatly below the assumed hydrostatic value. Proper account of this may change the temperature curve considerably, perhaps to something like that shown by the broken line in Fig. 66. At present it seems to the author that there is no evidence, one way or the other, as to the need for mechanical heating in the *non-magnetic* chromosphere.

Above the chromosphere lies the corona which, as shown in the next chapter, is at 1 000 000K or more. The corona is a poor radiator; but because of its high temperature and strong ionization it is a very good thermal conductor *along* lines of magnetic force. *Across* field lines it is a remarkably good thermal insulator. Therefore it conducts heat down towards the cooler chromosphere, following lines of force as they taper in towards the magnetic elements.

At lower temperatures, thermal conductivity drops rapidly, so that the temperature gradient becomes steeper and steeper in order to carry the heat. At the same time, the radiative emission increases at low temperatures, all the heat conducted downwards being radiated away just before

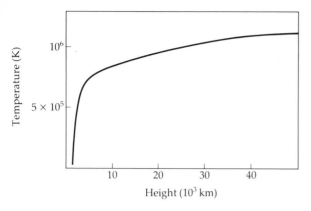

Fig. 67. The temperature in the transition region between the chromosphere (below 20 000K) and corona (above about 70 000K). The corona has a fairly uniform temperature because of its extremely high thermal conductivity, about 20 times that of copper.

reaching the chromosphere. The net result is a very sharp rise in temperature to 100 000K or more in only a few tens of kilometres (Fig. 67).

The region where the temperature rises rapidly forms the transition zone between the chromosphere and corona. It can be studied using far-ultraviolet spectral lines emitted selectively at different temperatures. All these lines are weak, and can be observed only with space telescopes (Fig. 68). In quiet regions, good agreement is found between the temperature in the transition zone and that expected from thermal conduction if the heat flowing down from the corona is about 2.5×10^{-3} Joules per square metre per second, but some additional heating may be needed in the lowest 20 km.

The density of the corona adjusts itself so that it and the transition region can just radiate away the heat generated there. This is achieved by a change in level of the base of the transition region. The corona is effectively in pressure equilibrium with the base of this region, and so with the top of the chromosphere. Hence a change in the height of the base of the transition region changes the coronal and transition region densities and emissivities. In this way a nice balance is maintained involving the heat generated, the thermal structure of the transition zone, the coronal and transition region densities, and the radiations emitted from them. If static, the base of the transition region would lie some 1500 km above the photosphere.

Now we turn to the real Sun, where motions play major roles. While waves can be observed in the non-magnetic chromosphere, they are much more significant in the magnetic component of the chromosphere. Here they are of two types: longitudinal – along the direction of the field lines; and transverse – at right angles to the field lines.

Longitudinal waves *inside* magnetic elements, mentioned in Chapter 6, affect the behaviour of the magnetic chromosphere, transition region and the corona beyond. Fig. 69 shows simultaneous velocity recordings of a wave packet at two slightly different heights. The wave period is nearly 5 minutes, and the oscillations at the higher level lag a few seconds behind the lower. When the difference in level is increased by using other lines, the lag increases to nearly 20 seconds. Although height differences inside the magnetic tubes are not well known, these observations show that a longitudinal wave is propagating upwards, and that the amplitude increases with height as the density falls. In Fig. 70, typical motions are shown *inside* the magnetic tubes and in the non-magnetic regions surrounding the tubes. It is easy to recognize the same disturbances inside and outside the tubes. The wave periods inside the tubes are sometimes well defined, sometimes erratic, but they are of the order of 5 minutes. These properties are the same as outside the tubes. The waves undoubtedly have identical origins, the oscillatory pressure fluctuations in or just below the photosphere described in Chapter 4. These squeeze gas up and down inside the magnetic elements.

The wave which propagates up into the

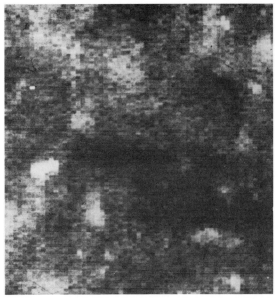

Fig. 68. *Skylab spectroheliograms of a portion of the Sun in (left) a carbon line emitted at about 55 000K (transition region) and (right) a magnesium line emitted at about 1 100 000K (corona). While the network has disappeared in the corona, some features of resemblance remain.*

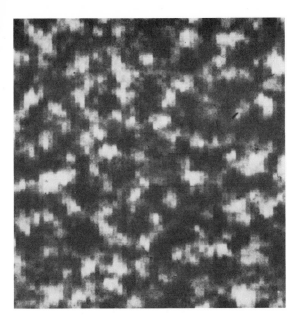

Fig. 69. *Vertical velocity variations inside a magnetic element as observed in two photospheric lines of iron, 5250 Å (from a lower level) and 5233 Å (from a higher level). These show a packet of almost identical oscillations having a period of about 5 minutes. The average disturbance at the higher level occurs a few seconds later than at the lower, showing that a wave is propagating upwards. The delay is much greater when these lines are compared with a chromospheric line originating at a much greater height.*

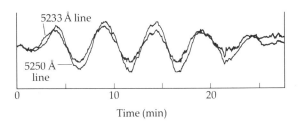

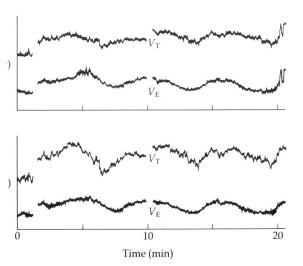

Fig. 70. *Vertical velocities observed simultaneously inside a magnetic element (V_T) and in the non-magnetic gas immediately outside (V_E) : (a) in the 5250 Å line and (b) the 5233 Å line. The gaps near 1 and 10 minutes are due to instrument adjustments which cause shifts in some or all of the traces. The traces are closely similar, showing that the oscillations inside the magnetic element and outside have similar origins.*

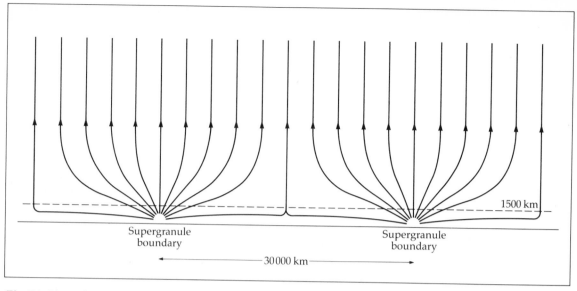

Fig. 71. Lines of magnetic force above an ideal supergranule. In the photosphere the fields are concentrated into the network axes, but spread out rapidly with height. It would be most unusual for the field structure to be as simple as shown here.

magnetic chromosphere is extraordinarily complicated. The gas is constrained to follow the field lines, and these open out with height almost radially to about 10 000 km from the network (Fig. 71). They then turn upwards into the corona, probably at quite an angle from the vertical. No complete, detailed theory of the waves is yet available, but we can appreciate some of their properties.

As the density drops with height, the wave amplitude increases so as to carry the same energy. There is a major difference between tubes of force which are nearly vertical or nearly horizontal. The former are relatively narrow at heights of 1500 km, and the wave amplitudes are correspondingly large. When the amplitudes become big enough, shock waves develop and cause heating, increasing the temperature and changing the height of the base of the transition region, pushing it up and down. The high velocities in spicules, up to 30 km/s, are due to these longitudinal waves. And it is also due to them that, at any one time, about equal numbers of fibrils (and spicules) have inwards and outwards motions such as seen in Fig. 56.

The amplitudes are much smaller at heights of 1500 km on the broader horizontal tubes, the fibrils, along which the gases oscillate backwards and forwards more gently.

An interesting consequence of the longitudinal waves is that the transition region must have an extremely fibrous structure. Waves in neighbouring tubes are independent, so that the base of the transition region is being pushed up in one, and falling down in another. The structure is highly dynamic and variable, but at any one time its appearance must be somewhat like Fig. 72, the base of the transition region being quite high along some tubes of force, very low along others.

Transverse (Alfvén) waves in Hα fibrils are observable with great ease. We need only obtain pairs of photographs of the chromosphere using light from opposite wings of the Hα line, and subtract one from the other as described earlier, to show the velocities. While they can be seen only when projected as a cinefilm, all fibrils on the Sun are found to be carrying Alfvén-type waves propagating mainly, perhaps solely, outwards from their roots at velocities

Fig. 72. Instantaneous locations of the bases of the transition region above a supergranule boundary. The base of the region, where the rapid increase to coronal temperatures occurs, oscillates along lines of magnetic force over height ranges of several thousand kilometres. The spicules are represented by the more vertical structures close to the supergranule boundary. The fibrils are the longer, more nearly horizontal structures.

typically 70 km/s. The sideways displacement velocity has an amplitude of some 5 km/s. This is so in both quiet and active regions, and even along Hα superpenumbral fibrils which extend out in radial or somewhat spiral-shaped forms beyond the penumbras of sunspots (Fig. 32).

At least two mechanisms contribute to the generation of the Alfvén waves. In the absence of a magnetic field, an ordinary longitudinal oscillation takes place along a straight line. If a magnetic field is present containing curved lines of force, the oscillating gas is forced to follow the bends. Action and reaction being equal and opposite, there is a corresponding force displacing the field sideways. This is greatest when the oscillatory velocity is highest, and zero when the oscillation is just about to reverse. At that time the field tends to return to its original position. The sequence of

back-and-forth displacements, which occurs progressively along a fibril as a longitudinal wave passes, sets up Alfvén waves propagating outwards. The second mechanism is the distortion of the magnetic canopy base by acoustic waves observed in the non-magnetic chromosphere. A third possibility is the direct sideways jostling of the filigree points, where the field emerges from the photosphere, by granule motions, but this seems to be less important than the other mechanisms unless the granules are deeper and more energetic than we believe at present.

The Alfvén waves are important sources of energy for heating the corona. The amount of energy required is still controversial, but the wave energy appears to be of the correct magnitude. The waves are seen propagating mostly, perhaps invariably, outwards along the Hα fibrils. As 90% or more of the field lines return to the Sun fairly soon rather than extend far out into interplanetary space, most of the energy must be dissipated on these lines above the highest levels at which Hα fibrils are observed, i.e. in the transition region or corona. Very little escapes to interplanetary space. Mechanisms of dissipation of Alfvén waves are not understood completely, though most astrophysicists are satisfied that no serious difficulty arises here.

Both Alfvén and longitudinal wave systems appear to have sufficient energy to account for coronal heating, and both are certainly important contributors.

9 Coming out from the Sun
– *the corona and beyond*

Fig. 73. The corona at sunspot maximum (eclipse of 16 February 1980). The photograph was taken through a radially graded filter to reduce the brightness of the inner corona and show the much fainter streamers of the outer corona.

The corona! The Sun's crown! The grandest sight of all! The inspiration for a delightful part of our cultural heritage! In widely separated parts of the world, Eurasia and the Americas, ancient people created virtually identical symbols of the Sun. The best known shows the solar disk surrounded by flames. It is still used widely by modern artists (Plate XI). A magnificent set of Sun-symbols appears in *The Sun in Art* (ed. Walter Herdeg, The Graphis Press, Zurich, 1962).

Fig. 74. The corona at the eclipse of 30 June 1973. This is more characteristic of the corona towards solar minimum, with long streamers extending from equatorial regions. The north pole (top) shows a fine set of polar plumes.

It is impossible to be certain about the origin of any prehistoric symbol, but the close similarity between the flame-type symbols in Eurasia and America suggests a common origin. Since the communities had been isolated from one another for tens of thousands of years, what could this be?

In a total solar eclipse, on average about once in 18 months, a narrow strip of the Earth's surface is shielded completely by the Moon from the brilliant disk of the Sun. The fainter corona then stands out in all its splendour. The disappearance of the Sun for a few minutes, although rare at any one part of the globe, must have created a tremendous impact on people ignorant of its cause. It is a strongly emotional experience even for us. What would they have seen?

The appearance of the corona varies greatly between the maximum and minimum of the solar cycle. At sunspot maximum, the brighter, inner parts form arches almost all around the Sun (Fig. 73). Surrounded by peaked structures like old-fashioned helmets, they taper off to long radial streamers further out. The same type of structure would be seen from all inhabited regions at some time or other. This seems almost certainly to be the common origin of the flame-like Sun-symbol.

What about the corona at sunspot minimum? Then the helmet-like arches disappear. Long streamers become more obvious, particularly in equatorial regions. Polar plumes, of which a fine example appears in Fig. 74, outline the polar unipolar magnetic field. Spoke-like streamers can be traced over distances of at least six solar diameters both at sunspot minimum and maximum (Fig. 75).

In Spanish caves and on rocks in Arizona, USA, are found almost identical paintings and petroglyphs which seem to be representations of the Sun and the coronal streamers (Plate XII). Sun-symbols in the present Huichol yarn paintings from north-west Mexico are direct descendants of the earlier petroglyphs (Plate XIII).

Many of the other Sun-symbols, ancient and modern, are variants on the above two forms. They all seem traceable to the appearance of the corona at eclipses.

The streamers and arches of Figs 74 and 75 lie along lines of magnetic force, as do the polar plumes. The cyclic changes of the corona are due directly to the changes in the Sun's field during the sunspot cycle.

The inner corona is about as bright as the full Moon, or as the sky on a clear day. It is no surprise that the corona is normally invisible beside the fiery sun. The brightness of the corona falls off rapidly away from the surface. Out to about half a solar radius, light is due mainly to the scattering of ordinary sunlight by electrons. Further out, light scattered by interplanetary dust predominates. Structureless, this can be subtracted in coronal photographs to reveal the outer streamer structure at its best.

That small, atomic-size particles scatter light effectively should be evident to us all, for skylight is simply sunlight scattered by molecules of nitrogen and oxygen in the air. There is, however, an interesting difference in that the sky shows the normal Fraunhofer absorption lines; apart from the component due to scattering by dust, the corona shows none! This is because the wavelength of the light scattered by any small particle is changed randomly by an amount of the order of the thermal Doppler shift. Scattering by electrons, the lightest and fastest particles of all, would smear out almost every Fraunhofer line, even at a temperature as low as in the photosphere.

About 1% of the light of the corona lies in emission lines. Until the identification in 1939 of two of these lines by Walter Grotrian in Berlin, the corona was thought to be at about the same temperature as the photosphere, or less. Grotrian discovered that these lines came from nine- and

Fig. 75. The outer corona at the eclipse of 16 February 1980. Many photographs of the same eclipse have been superimposed and processed so as to eliminate the huge brightness decrease away from the Sun. The long streamers then become very clear.

ten-times ionized iron atoms. Such highly ionized atoms can occur only where the temperature is extremely high. Within 3 years, most of the other coronal lines had been identified by the Swedish spectroscopist Bengt Edlén; they are due to atoms ionized 10 to 15 times, iron, nickel, calcium and even argon. For this, temperatures of 1 million to 2 million K were required. So the corona was much hotter than the photosphere and could not possibly be heated by solar radiation.

It was not long before other evidence was found of high coronal temperatures. Soon after the end of World War II, the new techniques of microwave radio were applied to study emission from the Sun. The high intensities of its radio noise confirmed that

temperatures upwards of 1 million K were present throughout the corona.

Radio waves longer than about 1 m in wavelength come only from the corona. In this respect, the radio emission is quite the reverse of the visible. The Culgoora radioheliograph, in Australia, provides images of the Sun which show how its radio emission extends much beyond the visible disk (Fig. 76a).

Fig. 76 also shows an X-ray picture (b). Extreme ultraviolet (EUV) and X-ray emissions became available for studying the Sun from the commencement of the space age around 1950. The rapid development of space technology led soon to a detailed knowledge of the spectrum, and later to

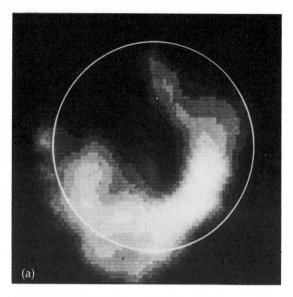
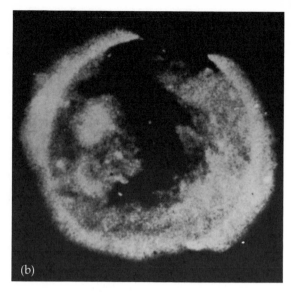

Fig. 76. *(a) 160 Megahertz (MHz) radioheliogram of the quiet Sun. Note that the disk of the Sun does not show as such on this image; the limb is superimposed artificially. (b) An X-ray image obtained at the same time has a very similar appearance. Both show a large coronal hole moving from the northern hemisphere into the southern.*

Fig. 77. *(right) An equatorial band of the Sun on 30 August 1973. (a) Magnetogram, positive fields bright and negative dark. (b) Ne VII spectroheliogram of the same region, originating at temperatures around 600 000K. (c) Similar Mg IX spectroheliogram at about 1 100 000K. (d) Fe XV spectroheliogram at about 2 500 000K. (b), (c) and (d) are all photographic negatives.*
Note that the fibrils show the magnetic field structure. The lines of force connect widely spaced active regions and other quieter areas of unipolar fields. The variations in intensity from line to line show that the temperature is much higher near the tops of the magnetic loops.

high-quality images either in broad spectral bands (X-rays) or monochromatic radiations. It so happens that, in the transition region from chromosphere to corona, where the temperature range is typically 10 000 to 1 million K, and in the corona itself, there are only fairly narrow temperature limits within which a given element can exist in a given stage of ionization. A balance is set up between the rate at which further ionization is produced by fast electrons which collide with atoms and ions, and the rate at which the various stages of ionization are reduced by recombinations between ions and electrons. In consequence, a given stage of ionization occurs in substantial quantities only near some particular temperature. For example, nine-times ionized iron indicates a regime around 1 million K. Further, the intensity of a spectral line can be used to find the thickness of the layer where such ions

occur. In this way, the temperature is found to rise very steeply from chromospheric values to 100 000 K, and then more slowly to the coronal 1 million K or more (see Fig. 67).

The great advantage of radioheliograms and monochromatic EUV solar images – spectroheliograms – in light from coronal emission lines is that they enable us to see the structure of the corona over the disk of the Sun as well as beyond; in this respect they differ entirely from eclipse photographs. Fig. 77 shows (top) a magnetogram with a sunspot of negative polarity (dark) just to right of centre, and some very small spots of both polarities a little further to the right and lower. Other strong magnetic concentrations are present. Below are spectroheliograms in lines of Ne VII (originating at a temperature of about 600 000K), Mg IX (1 100 000K), and Fe XV (2 500 000K). Fibrils are present with increasing strengths at ever higher

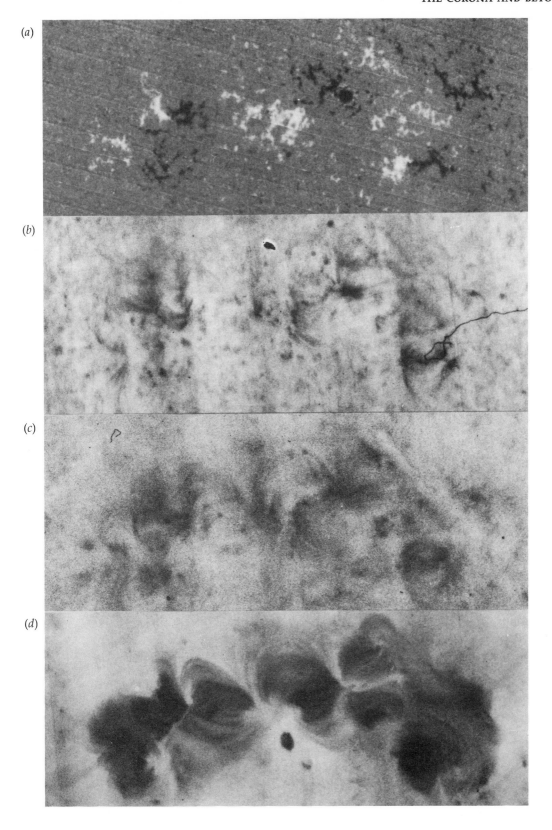

temperatures. They are aligned along lines of magnetic force which connect magnetic points far apart in the photosphere. The fibrils are shorter and incomplete in the Ne VII image, appearing only near the origins of the field lines. The Mg IX fibrils are continuous, while the Fe XV show the configuration best of all. Thus the tops of the loops seen strongly in Fe XV but not in Ne VII are hotter, and the lower parts of the loops are cooler.

The relation of the coronal field configuration to the photospheric field can be seen even more directly in the colour displays of Plates XIV and XV. The latter picture is rather confused beyond the limb because it overlaps other solar images, but we can get a good idea of the heights of the magnetic structures in yet another way.

Plate XVI is a set of X-ray images obtained on five consecutive days. Highly structured, the X-ray fibrils also show the magnetic field configuration. X-rays are emitted only by very hot regions. The dark regions are the coronal holes, seen side-on in Fig. 73. They are relatively cool, and coincide with large unipolar regions. There are usually coronal holes at the poles, except near sunspot maximum when polar field reversal is taking place. Other extensive coronal holes occur on the Sun, and these are carried east to west by solar rotation. From the edges of the coronal holes, arcades of magnetic loops, 200 000 km or so high, curve over into neighbouring regions. It is over these that the helmet structures of Fig. 73 lie. We shall learn in the next chapter why these and the streamers connect to interplanetary space.

Plate XVI shows another interesting

phenomenon, a random distribution of tiny X-ray bright points which last on average for only about 8 hours. They coincide with small magnetic dipoles which are also short-lived (Fig. 78). The dipoles are due to kinks which develop in twisted subsurface flux tubes, reconnecting, as described on p. 44, where the kinked tubes touch themselves. Thus a free flux loop forms, projecting through the surface. In a fairly short time, this floats up and away. It seems that the energy liberated during reconnection reappears somehow at the top of the loop to cause the heating responsible for the X-ray bright points. Since the Sun contains subsurface flux tubes at all stages of the cycle, X-ray bright points and ephemeral magnetic dipoles are found also at all phases of the cycle.

EUV radiation and X-rays have direct effects on the Earth, for they are absorbed in the Earth's upper atmosphere, ionizing the gases there and producing the layers of electrons that form the ionosphere at heights of 80 to 100 km upwards. These reflect radio waves efficiently. Without the ionosphere, long-distance radio communication would be impossible except by satellite.

The pattern of coronal magnetic fields discovered from EUV and X-ray images complements the picture of magnetic canopies derived in Chapter 6. The whole corona is filled with magnetic structures, and these can be traced down to the bases of the magnetic canopies, some few hundred kilometres above the Sun's surface.

But these canopies and coronal magnetic structures do not form instantly. When new flux is released by a new active region appearing at the surface, electric currents are induced in the highly conducting gases, and

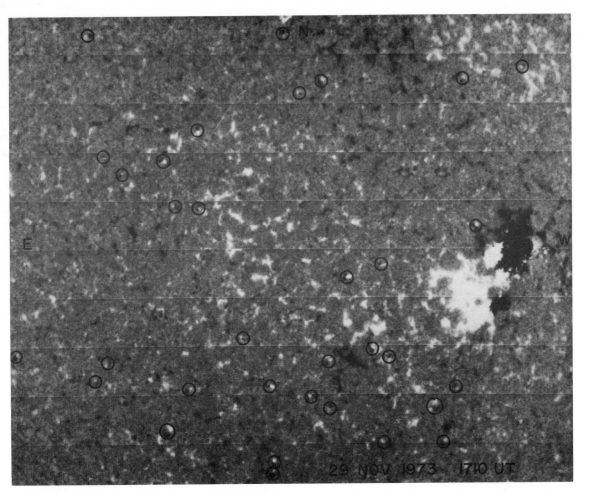

Fig. 78. Magnetogram showing ephemeral magnetic dipoles (encircled). These are features which lie below X-ray bright points.

these tend to resist changes in the magnetic field. Even so, reconnection of field lines to form more stable configurations takes place quite quickly – within some hours. This occurs by a complicated process known as the Petschek mechanism.

We dealt in Chapter 8 with the energy balance in the corona, pointing out that the source of heat was in part the dissipation of Alfvén waves, in part of longitudinal waves. In both cases the energy of oscillation propagates upwards from the photosphere along field lines. On the whole, about 10% of the Sun's flux passes into interplanetary space. About 90% forms well-defined loops or arches, some very low, some high but returning to the Sun after reaching no more

than about half a radius beyond the surface. Along these, energy is lost only by radiation, although the high thermal conductivity redistributes the heat downwards to cooler regions where the efficiency of emission is much enhanced. Thus most of the energy dissipated in the coronal loops is radiated away in the transition zone. Above coronal holes, where the field lines pass into interplanetary space, heat may flow outwards also, cooling the coronal gases there, but throughout the corona, the ultimate source of heat is the kinetic energy of turbulent motions in the convection zone. Without the convection zone, there would be no coronal heating, no corona.

And we would never have had the great

displays of the solar corona at total eclipses, or their injection into our heritage in the form of the ancient Sun-symbols.

Is the Sun surrounded by empty space? People thought so for a long time; but measurements from satellites show that the Earth is always surrounded by a stream of outflowing solar particles – the solar wind.

The existence and fundamental properties of the wind were predicted by Eugene Parker of the University of Chicago in 1958. He reasoned that, although the Sun and its atmosphere were held together firmly by gravitation, some of the particles would be fast enough to escape at the high temperature of the corona. The fastest are the electrons; but if they were to escape alone, the Sun would quickly be left with a positive charge which would stop any further loss. A charge does exist, just sufficient to make electrons and positive ions (mainly protons, i.e. hydrogen ions) escape in equal numbers. Quite near the Sun, the wind becomes supersonic, and continues supersonic further out.

Not long after Parker's theory was published, Mariner II observations showed that the wind was indeed present with speeds of the order of 400 km/s and particle concentrations near the Earth of some five per cubic centimetre, several million times less than in the corona. Both speed and density are variable, depending on coronal conditions.

We have become accustomed to the idea that magnetic fields decide the permitted directions of motions in the Sun's atmosphere. Then how do they interact with the wind?

If there were no wind, lines of magnetic force passing out through the photosphere would loop back into the Sun elsewhere. The strength and extent of the resulting magnetic field can be calculated readily from the observed photospheric fields. In particular, the further out from the surface, the smaller are the effects of the fine surface structures, and the field has a much smoother distribution out at some 1 to $1\frac{1}{2}$ solar radii (say 1 million km).

Large unipolar regions are clearly the ones which have greatest effect at these distances. Plate XVI shows the X-ray structure over a large unipolar region, appearing as a dark coronal hole bordered by bright loops where the field lines join neighbouring regions of opposite polarities. When a unipolar region is broad enough, its surplus flux has a major effect on the field at great heights. We may recall that the poles, too, are usually the sites of large unipolar regions and coronal holes. These also take part in forming the field at heights of 1 million km and more.

Without a solar wind, these large-scale fields would be stable, i.e. would stop any gas outflow, except where the gas pressure exceeded the magnetic pressure. Generally gas pressure is negligible everywhere in the chromosphere and the lower parts of the corona. However, at great heights, the rate at which the magnetic pressure falls off becomes greater than that for the hot coronal gases. Gas pressure then pushes the field out further and further, until continuous flow occurs. In the rare case where the only unipolar regions are at the poles, the field

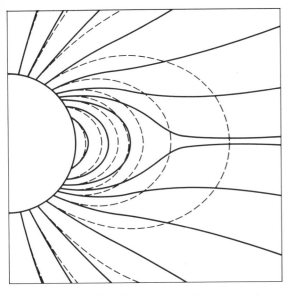

Fig. 79. Magnetic field lines near the Sun when polar regions are the dominant source of magnetic flux (section through polar axis). Broken curves show what the field lines would be in the absence of coronal gas; continuous curves show the actual field lines as blown out by the solar wind.

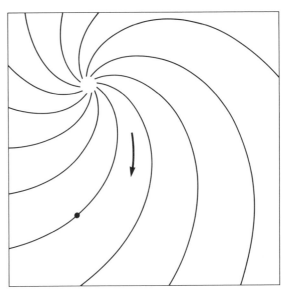

Fig. 80. Spiral shape of the interplanetary field. The arrow shows the direction of gas motion outwards from the Sun in the solar wind. The black dot represents the Earth.

lines would be drawn out as in Fig. 79. The equatorial plane separates outward and inward field lines, and the wind can flow out along all field lines open to interplanetary space, but now the weak field must conform to the direction of motion of the gas.

Usually, low-latitude flux complicates this simple structure, even at sunspot minimum. Near the boundary between unipolar regions of opposite polarities, the field is strong enough to form lower-altitude loops between these regions. Higher, the wind dominates and blows out the overlying helmet arches.

The wind and field have many important properties. At the distance of the Earth, the field is typically 5×10^{-5} gauss, just over one ten-thousandth of the strength of the Earth's surface field at the equator. However, the geomagnetic field drops off rapidly with height, and equals the interplanetary field about 10 to 12 Earth radii out. Here complex interactions occur, enabling solar gases to enter the Earth's magnetosphere. The fast solar particles then follow lines of force down to the Earth's polar regions, producing aurorae when they collide with and excite terrestrial atoms at altitudes above 100 km.

Since the Sun is rotating, the particles in the wind also have velocity components of 4 km/s in the direction of rotation at two solar radii ($2 R_{\odot}$) from the solar axis. They retain this velocity component, which is quite small by comparison with the average wind velocity, as they move outwards. But at increasing distances, the corresponding angular velocity drops, so that the particles lag behind the Sun. They move in spiral paths, and force the interplanetary field to be spiral-shaped. Continuing out beyond the Earth, they make the field almost circular at the distance of Jupiter, although the rate of outflow of particles is constant.

From what part of the Sun do the lines of force come that connect with the Earth? At a typical speed, 400 km/s, it takes rather more than 4 days for the wind to flow from Sun to Earth, during which the Sun rotates through about 55°. By the time particles reach Earth distance, their origin has rotated about 55° westwards, so that wind particles reaching the Earth originate from a region now 55° west on the Sun. Fig. 80 shows typical field lines, and the relative locations of Earth and the point of solar origin.

Now a subtle but large-scale effect becomes apparent. Fig. 79 showed the field when flux originated near the poles alone. What happens when fields closer to the equator are taken into account? If, for example, a coronal hole extended from equatorial to polar latitudes, it would be equivalent to extending the polar field towards the equator in that neighbourhood, distorting the surface which separates outward and inward interplanetary fields and pushing it towards the opposite pole. Calculation and observation show that this happens usually twice around the Sun. The structure is preserved by the wind beyond a radius or so out, so that the surface separating outward and inward fields wanders regularly in solar latitude, making usually two oscillations around the Sun. In interplanetary space the surface of separation is no longer plane but warped (Plate XVII).

This property was discovered from satellite observations. Four times each 27 days (a solar rotation), the interplanetary field usually reverses direction suddenly. This must happen when the satellite – and the Earth – pass through the surface separating the fields, in the zone of the warps. The regions of uniform field direction are known as sectors, and are surprisingly constant in solar longitude when mapped back onto the Sun. This shows that, in a way not yet understood, the large-scale magnetic fields in sunspot zones have long-term stability over durations of several years.

10 Beauty and action
– *prominences*

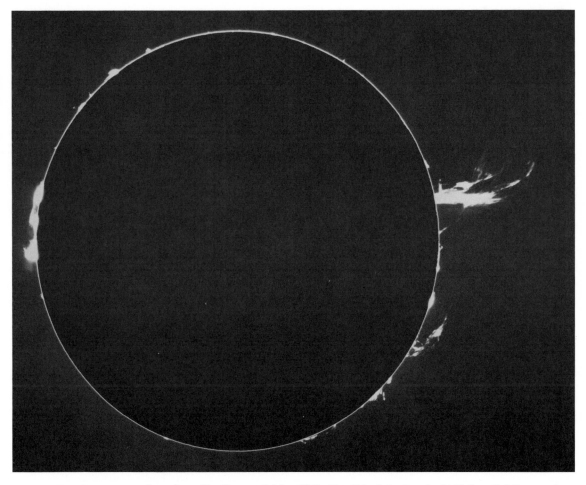

Fig. 81. *Prominences seen through an Hα filter on 7 May 1971. The disk of the Sun is shielded artificially to reduce glare. A huge, short-lived ejection (a 'tennis racquet' structure) has been thrown out by a flare. Below is a fine example of a coronal cloud, out of which gases stream downwards along slender, curved trajectories. A large cap prominence is present on the opposite limb.*

The spectroheliogram of Fig. 54 shows long dark markings scattered over the Sun at sunspot latitudes and higher. Known as filaments when on the disk, they appear very different when solar rotation carries them to the limb. Bright against the dark background of space and extending typically 40 Mm outwards (about 1/20 of the solar radius), they are then called prominences (Fig. 81). Brilliant magenta objects at total eclipses,

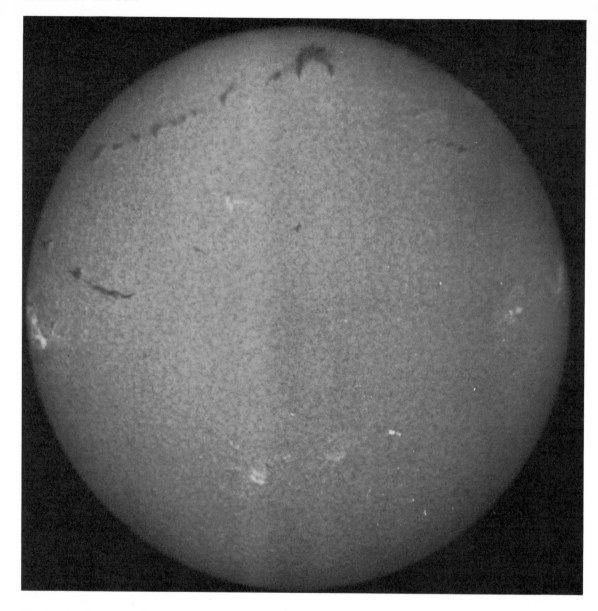

Fig. 82. A polar crown of prominences (4 October 1966).

their light comes mainly from the red Hα and blue Hβ spectral lines of hydrogen, and the violet H and K lines of ionized calcium. They are usually observed through Hα filters, appearing as giant flames rising from the Sun's rim.

The first great study of prominences was carried out by the French husband-and-wife team M and Mme Lucien d'Azambuja. In 1948 they were able to announce that at least one-third of all prominences are born in spot groups about 3 weeks or so after the spots (for good examples see Figs 58, 62 and 64). They extend gradually towards higher latitudes, tilting continuously away from the north–south direction. After the decay of the group, they continue to move polewards, becoming almost east–west at latitudes somewhere in the range 35° to 65°. Here they lose their identities and, with other like

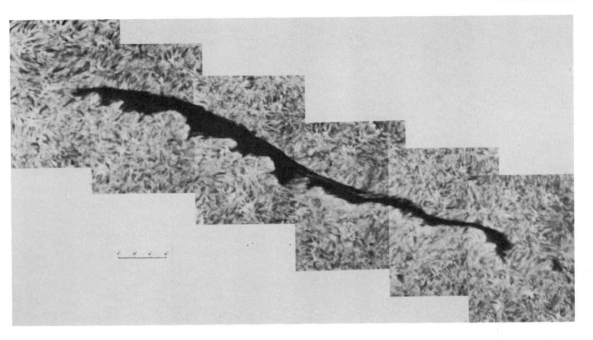

Fig. 83. A quiescent Hα filament on the disk: 520 000 km long and over 40 000 km high, projections towards the centre of the disk (below) are the footpoints where the prominence is connected to or towards the chromosphere. The fine structure is diagonal, showing that the magnetic field direction is skew. The fainter parts of prominences seen at the limb are invisible on the disk.

prominences, form an intriguing 'polar crown' (Fig. 82). Prominences born outside spot groups have indistinguishable behaviours. Lives of individual objects are typically 4 to 5 months.

By 1954, Horace and Harold Babcock had discovered that filaments lie along lines separating regions of opposite magnetic polarities – along the axes of the coronal helmet structures and below the helmets. Their poleward drifts and inclinations are exactly those expected from the combination of the poleward flow of gases, such as observed by Duvall, and differential rotation. We may look at a magnetogram such as Fig. 42 or 44 (b) and locate the possible filament sites on it very easily.

Prominences are of various types, but we can separate them into quiescent (lying in quiet regions) or active region prominences. All have delicate internal structure and some degree of motion.

Fig. 83 is a mosaic of Hα filtergrams showing the structure of a quiescent filament. The further side is almost smooth,

the nearer side has large irregular projections. The interpretation is that most of the filament lies above the chromosphere, extending down towards it only in isolated places.

It is often said that the projections of the filament towards the observer represent the footpoints of magnetic arches. In this case we should be able to find magnetic elements underneath. Attempts to discover these have been frustrating failures. Whether the quality of the observations is inadequate or the general concept incomplete is uncertain. Perhaps the lines of force extend much further than shown in Hα. We can be quite sure that prominences are *not* the result of abnormally strong fields.

When seen at the limb, many (perhaps all) quiescent prominences have structures such as in Fig. 84. There is an array of almost vertical ropes down which gases stream quite slowly. With diffuse upper ends, the lower ends terminate abruptly on curved arches, obviously magnetic. It is by no means obvious just how such limb appearances fit

91

with the structures seen on the disk in Fig. 83. There is no possibility of understanding the origin and condensation of quiescent prominence material, its support and motions until a satisfactory three-dimensional geometrical model is available embodying the structures seen on the disk and at the limb. Strangely, no proper observations have yet been made of prominence geometry, as if astronomers were just not interested in solving such problems. Nor has any attempt been made to examine the magnetic field directions in the extraordinary fine structures present.

Yet we can make some progress. To emit Hα and the calcium H and K lines strongly, a prominence must be cooler than 10 000K. Otherwise there would be too few neutral hydrogen and once-ionized calcium atoms; but the cool prominence sits happily in the 1 million K corona for months and months. To make things even worse, Plate XVI shows, particularly near the boundaries of coronal holes, that magnetic fields containing *very* hot gases loop over from one polarity to the other. The prominences lie under these loops.

To resolve this dilemma, we need recall that the flow of heat across a magnetic field is very slow, whereas coronal gases conduct heat well along field lines. This simplifies things greatly since prominences are then insulated on their extensive sides and can be exposed to conductive heating only along their narrow cross-sections. The problem cannot be made to vanish by having low temperatures all along the field lines connecting the prominence to the chromosphere. At a temperature of say 10 000K, the drop in pressure over a height range of 30 Mm would be far too great. The drop is vastly less in the 1 million K corona – and measurements show that gas pressures in prominences and the surrounding corona are identical. Therefore, the prominence must be open to gas flow and thermal flow from the corona along the field lines.

The survival of cool material heated by thermal conduction along magnetic field lines – probably from both ends – is just like the problem of the chromosphere–corona transition region (Chapter 8). Although the conductivity is high at 1 million K or more, it drops rapidly at lower temperatures. Thus the cooler parts of prominences are good insulators, even along field lines. As in the chromosphere–corona transition region, the emissivity increases greatly as the temperature falls, so that conducted heat is radiated away increasingly at lower temperatures. Calculations show that cool prominence material heated by conduction can exist very nicely as long as there are transition zones to the corona at both ends of the field lines. Away from the prominence, the temperature rises to 100 000K in a few tens of kilometres. Beyond this, there is a slower, more extensive rise to coronal values in perhaps 10 Mm or so.

Observations of this type of thermal structure surrounding quiescent prominences have been made from rockets during eclipses. Sufficient detail has not been available to provide full proof because of our lack of knowledge of prominence geometry, but the results are consistent with predictions as far as they go.

The magnetic fields are stronger close to active regions, and prominences there have different appearances and behaviours with

Fig. 84. A quiescent prominence at the limb. It consists of an array of fine, almost vertical ropes with diffuse upper ends. Gases flow downwards at speeds of about 1 km/s, but the random motions are somewhat faster.

rapid internal motions. Dense condensations flow down the legs of fine, curved arches at speeds up to 50 km/s or more. Presumably they do so under gravitational attraction, guided by the strong magnetic field. The detailed origin of the gas involved is once again unknown because of our ignorance of the magnetic field structure.

Rather like the snake that sloughs off its skin to grow, prominences pass through the occasional traumas of eruptions during which they blow off from the Sun in the course of a few hours ('sudden disappearances'). Fig. 85 is a sequence of Hα filtergrams showing a prominence, originally on the disk. On March 11 and 12, 1970, its behaviour was quite normal, but the following day it commenced rising. Within 2 hours it developed into a great eruptive prominence, forming an arch extending far beyond the limb. In the meantime, the filament vanished from the disk. Fig. 86 shows another eruptive prominence.

The interpretation of such events in terms of internal structure is debatable, though cinefilms seem to show that individual prominence strands are gently twisted. The prominence disappears from the Sun completely. Much of its gas is ejected from the Sun, but some flows down at the outside ends of the arch.

These sudden prominence disappearances are not uncommon. Between March 11 and 12, the big filament on the right-hand side of Fig. 85a vanished also, without trace.

Following two-thirds of these eruptions, the prominences re-form a few days later in much the same positions, as if nothing had happened! Eruptions are part of the normal prominence life-cycle.

How can we account for this

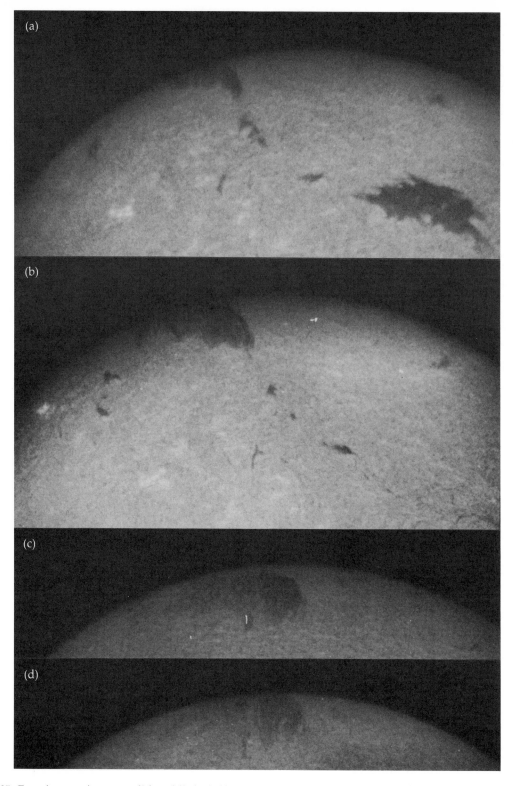

Fig. 85. *Eruptive prominence on disk and limb. (a,b) Appearance on 11 and 12 March 1970 before eruption. (c,d) 13 March 1970 at times 1817, when the eruption was just commencing, and 1912. (e,f) Limb appearances at times*

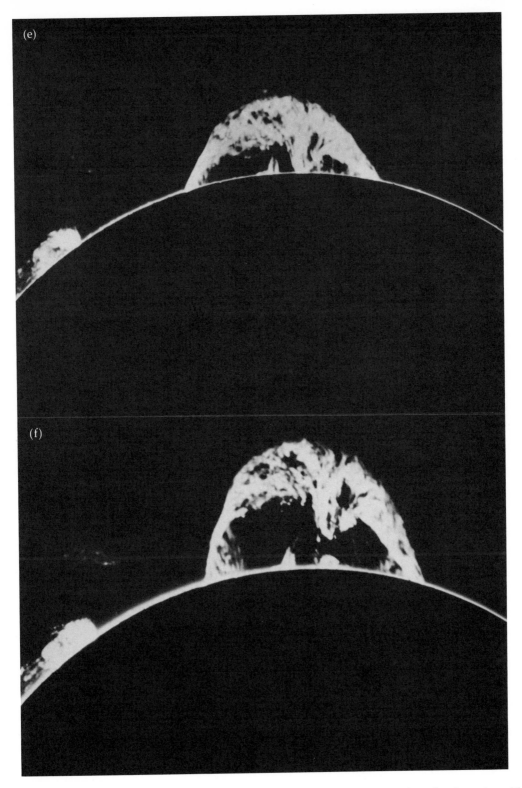

1959 and 2010. The base of the prominence as seen at the limb is already 110 000 km above the photosphere. The full height in (f) is 370 000 km, and it is rising at 160 km/s.

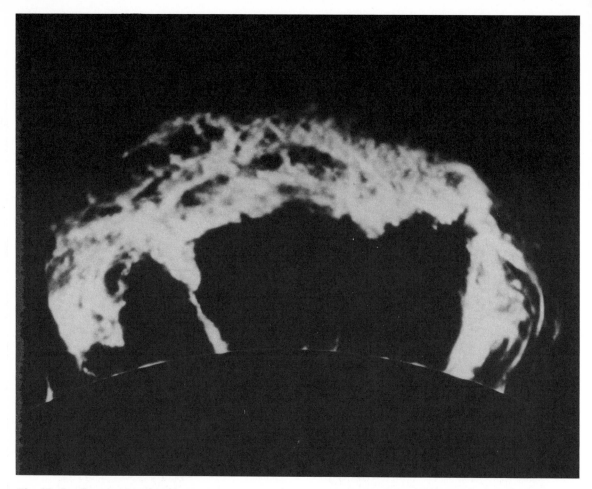

Fig. 86. Another eruptive prominence.

extraordinary behaviour? The phenomenon can only be magnetic. Shearing of the field lines by differential rotation increases the magnetic energy greatly. If the lines of force were able to reconnect, the magnetic energy would be reduced and the process would proceed. This would happen instantly if no highly conducting gases were present, but if adjacent field lines are parallel locally, as in a simple twisted tube, no reconnection process such as Petschek's mechanism is available.

Beyond this, all is speculation. However, if tubes of force are gently twisted, as all evidence suggests, *adjacent* tubes have field lines which are no longer parallel where they are pushed against one another, but cross at appreciable angles. Then it might become possible for reconnection to occur. This would produce an array of shortened tubes and a longer arch-shaped tube connecting the furthermost footpoints. The huge arch would expand upwards and cause the ejection of prominence material.

There may well be other possible triggers for the reconnection of field lines. Further observation and study will be needed to show what the cause is. More hard work, but it will be worth it. They *are* grand events!

11 Action and beauty
– *flares*

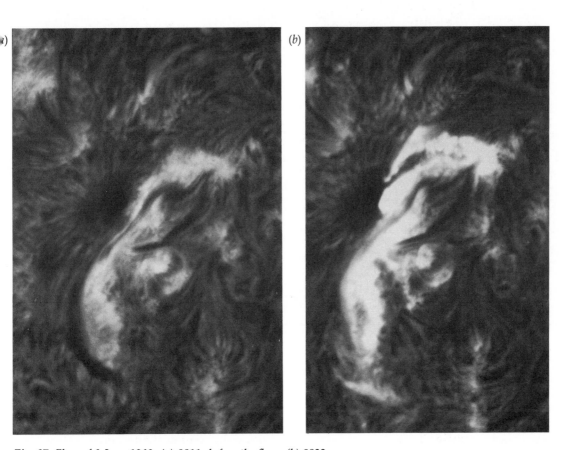

Fig. 87. *Flare of 9 June 1968. (a) 0011, before the flare. (b) 0032.*

Flares are Happenings. As late as 1960, we would have called them chromospheric flares. Small flares may last only a few minutes, big flares an hour or more.

A rather nice flare of medium size is shown in Fig. 87. Photographed through an Hα filter, it developed gradually between 0020 and 0032 on 9 June 1968 without warning of the drama to follow. Then for 8 minutes, it expanded suddenly at 100 km/s. At the same time, a pre-existing filament disappeared and typical flare ribbons

developed on either side, gradually moving apart. All parts of the flare consisted of very fine filamentary structures. Later, dark loops lay over the decaying flare not too far from the sunspot. In some flares these loops become very pronounced (Fig. 88).

The flare of 9 June 1968 occupied 450 millionths of the Sun's disk. A huge flare, about four times larger, occurred in another spot group a few hours later. Flares larger than this are rare, only one or two during a cycle. As many as 25 small flares can be

(c)   (d)

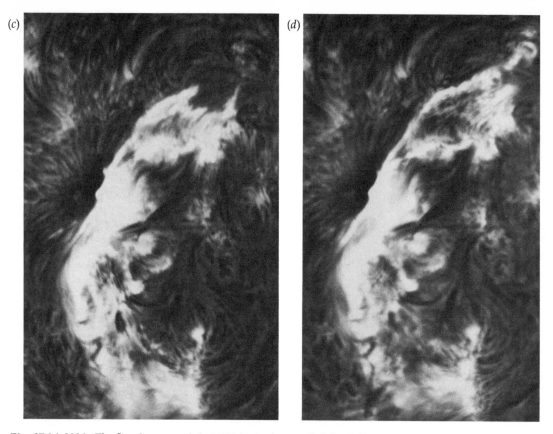

Fig. 87.(c) 0038. The flare has expanded at 100 km/s along well-defined filamentary structures at its upper end, and extended more diffusely near the bottom of the frame. (d) 0039.

recorded daily around sunspot maximum. There are very few at sunspot minimum. Flares are of varying forms and shapes, but most occur near spot groups. The few that don't are usually associated with disappearing filaments, and may be quite large.

Over the past 30 years, measurements of radiations in other parts of the spectrum have shown that the Hα or chromospheric flare is only a small part of the phenomenon (Fig. 89). Often, the first hint of a flare is found in the darkening and rapid up-and-down motions of filaments. Perhaps 20 minutes later, a flare precursor begins slowly with the emission of soft X-rays and short-wave ultraviolet, followed soon afterwards by Hα. The seat of the flare is where the X-rays and ultraviolet are emitted, in the corona. About 5 minutes into the flare, there is a sudden impulse of hard and

medium X-rays, and bursts of microwave radiation which also come from the corona. Even after this, in the main phase of the flare, emission continues to increase in most parts of the spectrum, decaying gradually after a rather gentle maximum.

Microwave observations can show flares at coronal levels with high time resolution. Figure 90a is an example of a type II radio burst following a flare. The spectrum consists of two pairs of frequencies, one pair being at just twice the frequency of the other. This relation, in which the higher frequency is known as the second harmonic, shows that there is a direct physical connection between them. All frequencies decrease with time.

What is the interpretation of this curious phenomenon? A highly-ionized gas, or plasma, has an additional non-acoustic way of oscillating longitudinally which has some

(f)

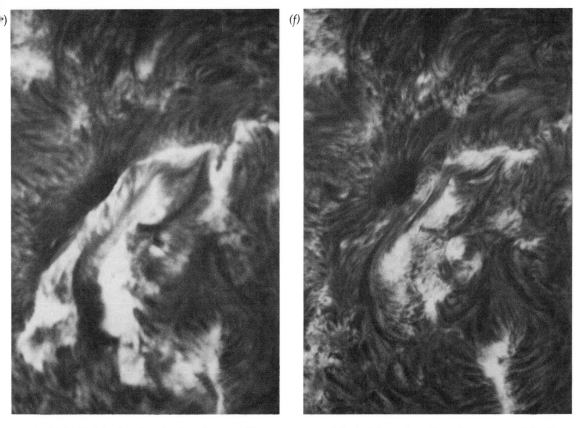

Fig. 87.(e) 0107. The flare has developed a two-ribbon structure, while dark loops have formed over parts of the flare near the sunspot. (f) 0234. After the flare. The structures are surprisingly like those in (a), although differences can be seen to the right of the sunspot.

resemblances to sound. When undisturbed, each tiny volume is electrically neutral. If the extremely light electrons in some region are all displaced with respect to the much more massive ions, powerful electrical forces are set up and pull the electrons back. In the absence of friction, they overshoot and are pulled back again from the opposite direction, oscillating at a radio frequency which depends only on the electron concentration and falls off slowly as the latter drops. This frequency is known as the resonant plasma frequency, or simply the plasma frequency, and the oscillations can result in the emission of plasma radiation at the same frequency.

While the cause of the splitting of the spectrum of type II bursts into a pair of frequencies is controversial (and splitting is by no means present in all flares), the height

of origin of a given frequency is in most cases that where the corona has the same resonant plasma frequency. Thus, from the known properties of the corona, the frequency gives the height. In Fig. 90(a) this increases from 1 080 000 to 1 430 000 km in 5 minutes, indicating that the corona was being excited by something ejected through it at a velocity of over 1100 km/s.

The Culgoora (Australia) radioheliograph, which operates at several fixed frequencies, can be used to locate microwave sources. The most dramatic ejection ever recorded (Fig. 91) was of a moving type IV event, on 1 and 2 March 1969 (the special features of these events, which have much in common with type II bursts, will be described shortly). It could be traced out to 5¼ solar radii beyond the limb. Observations slightly earlier at Hawaii

(a) (b)

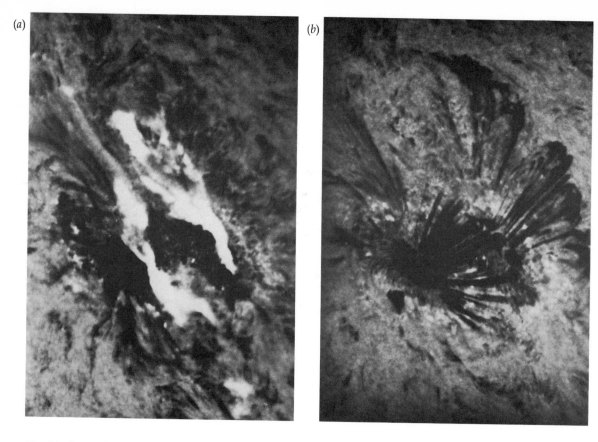

Fig. 88. *A two-ribbon flare (a) by the end of which an extensive loop system has developed (b). The loops are extremely hot at their tops, and gases flow down both sides of each loop at velocities up to 150 km/s. They can last for several hours.*

showed a high-speed Hα prominence ejection originating from the same part of the solar limb. This and almost all type II events are produced by flares, although the flare may be behind the limb and hence invisible. In this way, we can track, for the first few per cent of their paths, the solar particles which are ejected into interplanetary space. These produce geomagnetic storms and aurorae on Earth if the particle stream is aiming in the right direction.

Radio frequencies *can* be excited in ways other than plasma emission. One is when electrons gyrate around lines of force in a magnetic field. At times, this can serve as a neat way of studying magnetic fields at considerable distances out into the corona. For example, Plate XVIII shows such emission from a magnetic arch which rose by

more than a solar radius within half an hour, at a speed in excess of 400 km/s. The colours mark where lines of force in the arch have components towards us (e.g. red) or away from us (e.g. blue).

The microwave spectrum often shows extraordinarily rapid changes (Fig. 90(b)). Known as type III bursts, these can be interpreted only as the ejection of electrons at high velocities of the order of three-tenths the speed of light. They are coincident with rapid increases in the intensity of a flare or part of a flare. The radioheliograph gives the location of type III bursts in space. For example, Plate XIX (*top*) shows the emission at three fixed frequencies as the ejection passes through. As expected, the measured heights correspond to emission at the plasma frequencies.

Fig. 89. Emissions by a typical flare in different parts of the spectrum. The occurrence of radio bursts is indicated at the top of the chart, and the ejection of high-energy electrons and protons at the bottom.

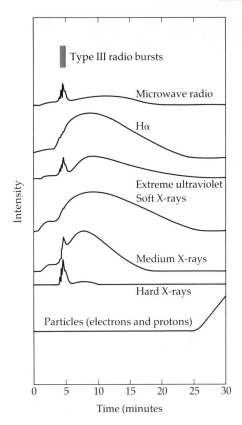

Type III radio bursts

Microwave radio

Hα

Extreme ultraviolet
Soft X-rays

Medium X-rays

Hard X-rays

Particles (electrons and protons)

Intensity

Time (minutes)

Plate XIX (*bottom*) shows a different type of microwave burst, type IV, where all frequencies originate in the same place. This is clearly not due to emission at the plasma frequencies, which would be at different levels. It is produced more likely by synchrotron emission from electrons. Type IV bursts are related more closely to type II than to type III, and they, too, are produced by flares.

The great extent of disturbances produced by flares is illustrated in Fig. 92. Due to a flare behind the limb, the corona was excited to heights of a solar radius or more over an arc covering 180° in latitude.

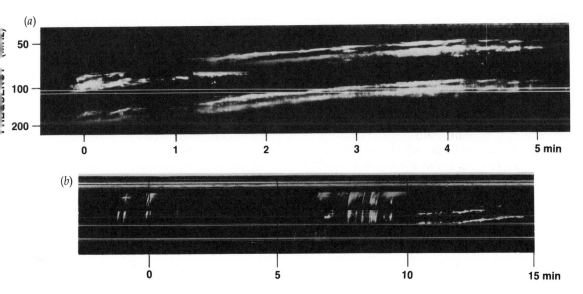

*Fig. 90.(a) A type II radio burst following a flare. The spectrum is plotted on the vertical axis and its time variation from left to right. The spectrum consists of two pairs of frequencies, the lower being the second harmonic of the upper. The drift in frequency with time is due to the passage of particles ejected at a speed of over 1100 km/s.
(b) A set of type III bursts due to the ejection of electrons at velocities of the order of one-third that of light. Type III bursts are produced by rapid changes in intensity in parts of a flare. A type II burst can be seen near the right-hand end of the figure.*

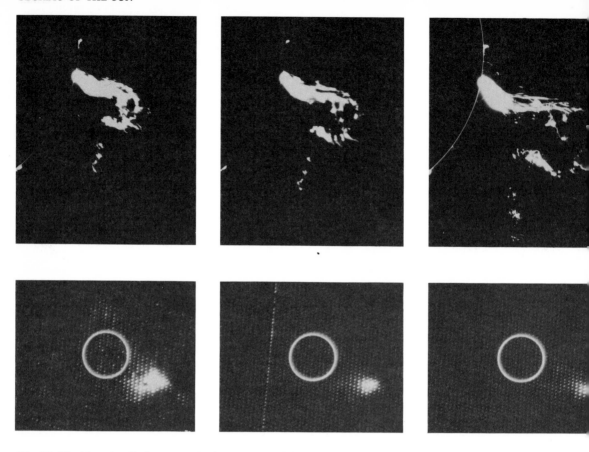

Fig. 91. The biggest radio-frequency ejection ever recorded, a moving type IV burst (1 and 2 March 1969). Above, as observed optically in Hα; below, the appearance at the radio frequency of 80 MHz. There is a large change in scale between upper and lower sequences.

Only rarely are flares seen in white light, yet a white-light flare was the first ever observed, on 1 September 1859. Richard Carrington, manager of the Brentford brewery and a keen astronomer with a private observatory at Redhill, Surrey, and, independently, amateur R. Hodgson at Highgate, in London, were both excited to find a small, extremely brilliant white patch near a large spot group. After about 5 minutes it disappeared, to be followed about 17 hours later by a mighty magnetic storm, one of the greatest in over 100 years, and a magnificent aurora. These chance observations caught a flare probably as large as has ever occurred. Beginner's luck? The largest prominence was recorded shortly after the Sacramento Peak 'Hα coronagraph' was commissioned!

How big, how energetic can flares be? A very large flare (Fig. 93) emits a great deal of radiation during its life of an hour or so. An even bigger amount goes into heating particles and ejecting them into the interplanetary medium. How can we appreciate the total energy involved? Suppose there were a huge fire on Earth. There isn't enough combustible material to combine with all the oxygen in the atmosphere. But just suppose there were, and that this huge fire used up all the oxygen. No further burning could take place. The energy produced would be about one-sixth of that in a very large flare. A rather impressive event and we can watch it all!

X-ray photographs often show that the hottest parts of flares are extremely small,

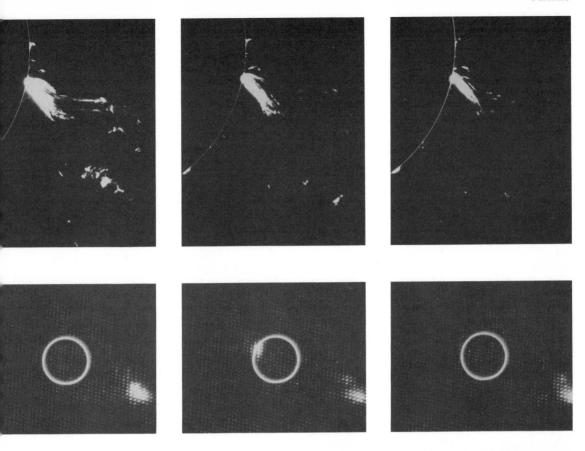

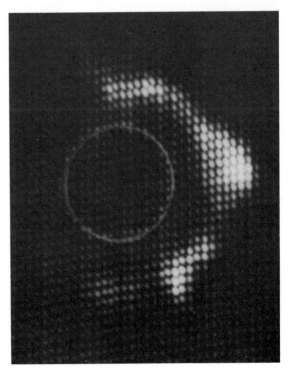

Fig. 92. (right) Coronal radio emission excited by a flare behind the limb.

below the 3500 km limit of resolution. Spectral lines of 24-times ionized iron (Fe XXV) are found, requiring temperatures above 20 million K, perhaps up to 100 million K. At times, soft X-ray emission is located in larger loop systems, the hottest parts being at the tops of the loops. Decay stages of large flares show Hα emission developing in loops. Loops are also seen in spectral lines, e.g. from Fe XIV originating at 2 million K, and in microwave images as in Plate XVIII. Obviously the loops are not the same in Hα and Fe XIV, but the total of all observations leads to the conclusion that loops are basic to flares.

Where does the energy come from? What causes flares? Such a huge amount of energy cannot be transported to the flare site in the short time available, and so it must be

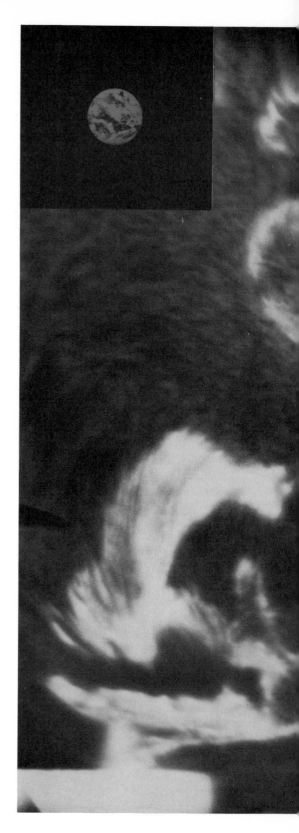

Fig. 93. A very large and bright flare showing two-ribbon characteristics. The frame is 220 000 km across. The insert shows the Earth to the same scale.

present locally. Astronomers are agreed that it can only be in the form of magnetic energy, in the excess energy that exists when field lines are distorted. Just as in eruptive prominences, to which many flares are closely related and which are often associated with large two-ribbon flares, the field lines are distorted by the motions of their footpoints at photospheric level. Once again, the field lines are unable to reconnect and release energy until some suitable disturbance takes place. However, in no single case have we been able to trace the sequence of magnetic events in detail.

This doesn't stop enthusiasts from guessing at what the magnetic configurations might be. There is little doubt, for example, that Petschek's mechanism, or some variant of it, is involved in the impulsive phase.

For the main phase of a flare, American astronomers Roger Kopp and Gerald Pneuman have suggested a mechanism involving the excitation of loops at successively greater heights. The loops appear when reconnection occurs between outward and inward field lines, just as in eruptive prominences (Fig. 94). The hottest part is where the energy is released and outward-flowing gases are stopped suddenly, at the top of the most recently formed loop. Lower loops cool partly by radiation, partly by conduction. As in the normal transition zone, it becomes a neat thermal problem to decide just what process dominates. Under some conditions the temperature can be low enough for neutral hydrogen atoms to appear, giving rise to Hα emission or absorption.

What about the *chromospheric* flare? This is due to hot electrons diffusing down lines

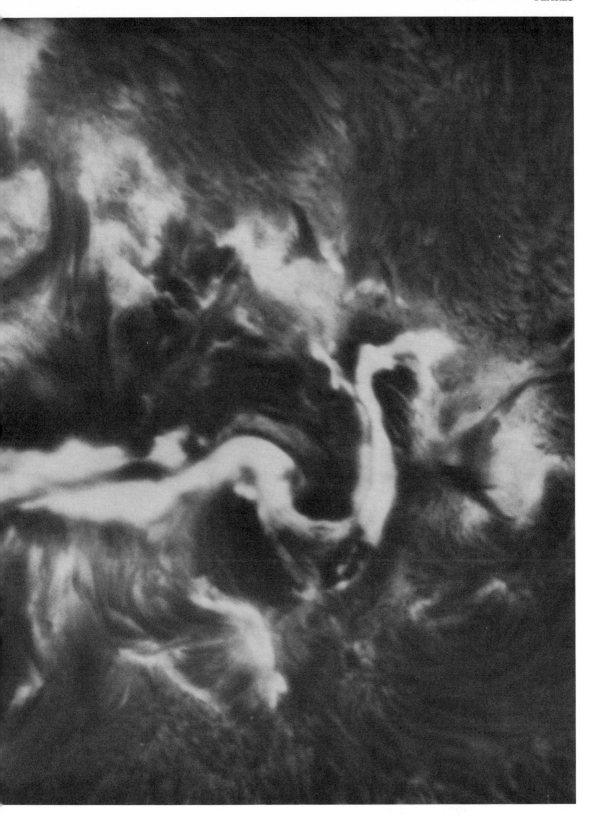

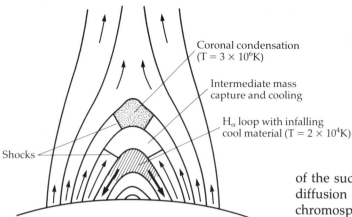

Coronal condensation
(T = 3×10^6K)

Intermediate mass
capture and cooling

H_α loop with infalling
cool material (T = 2×10^4K)

Shocks

Fig. 94. The formation of flare loops according to American astronomers Roger Kopp and Gerald Pneuman. Lines of force, directed outwards on one side of the system and inwards on the other, reconnect in the corona to form loops. The direction of gas flow is shown by the arrows. Shock waves form where outward-flowing gas is suddenly stopped, at the top of the most recently formed loop. These, the hottest parts, cool gradually and fall back into the chromosphere.

of magnetic force, colliding with chromospheric hydrogen atoms and exciting them directly or ionizing them. In the latter case, flare radiation is emitted as the electrons and hydrogen ions recombine. White-light flares occur in those rare cases where the diffusing electrons have sufficient energy to penetrate the chromosphere and heat the photosphere.

What caused the sudden expansion of the 9 June 1968 flare between 0032 and 0040? It is strange that we could have studied this better with the instruments of 40 years ago than today! In our feverish desire for automation, we have lost the ability to measure Doppler shifts in flares as could be done readily with the simple Hale spectrohelioscopes used visually in pre-War days. In consequence it is impossible to say *from observation* whether the chromospheric gases themselves were moving at such high speeds as 100 km/s. It is possible that they were, for we have many observations of prominence ejections at such speeds, and even fast-moving ejections of flares. Most astronomers would probably prefer to think

of the sudden flare expansion as due to the diffusion of hot electrons through the dense chromospheric matter, but we really need direct observations to be certain.

Flares emit huge amounts of radiation which ionizes atoms normally present in the Earth's upper atmosphere. The ionosphere is greatly disturbed at such times. The propagation of short-wave radio waves is then blocked by the appearance of a radio-absorbing layer at a height of about 80 km, causing short-wave radio fadeouts.

Flares also eject huge numbers of particles into the interplanetary medium. These travel along field lines, often perturbing them in the process. On reaching the Earth, they produce aurorae when electrons collide with and excite atmospheric atoms. In a complicated way, they also produce electric currents at great heights, causing major, though temporary, changes to the Earth's magnetic field. On average, about ten magnetic storms are produced each year, more near sunspot maximum when flares are most frequent. About one-third of these are associated with flares, commencing about a day or so after the flare. The delay is due to the time required for particles to travel from the Sun to the Earth. Those not associated with flares form members of a series recurring at 27-day intervals. They occur because the large, long-lived unipolar regions coincide with coronal holes, from which the field lines extend far out into interplanetary space. These are the places of origin of the particles which cause recurrent geomagnetic storms as the appropriate lines of force are carried across the Earth by solar rotation, once every 27 days.

12 Sunspots and the weather

Do sunspots affect the weather? Many people have said so, but it isn't easy to prove. Queensland farmers and graziers think that it is worth paying for weather forecasts based on sunspots, but this is not a reliable test. Until recently, scientists had an almost united answer – No! However, a select committee of the Australian Academy of Science pointed out in 1976 that solar activity might well affect weather, but if so the effects must vary from one geographical location to another. No mechanism is known for any such connection.

Weather is chancy. Weather is different in different places. A big sunspot is bound to coincide with some major weather change somewhere. We remember the coincidences and forget about the rest. There needs to be a way of testing whether the coincidences occur only at chance rates, or whether they are more frequent. A very extensive branch of mathematics – statistics – deals with such matters. Statistical tests giving the significance of coincidences are often easy to apply, but there are some cases where they are very difficult. Despite recent progress, for example, there is still much controversy between scientists as to how to tackle the significance of coincidences in time. While controversy is healthy, and stimulates improvements to methods of analysis, it is aggravating because there are even now no methods of test which are accepted universally by scientists.

In this chapter we shall come to a very definite, positive conclusion about the effect of solar activity on weather, but we must avoid all analyses about which there can be serious doubts.

Sunspots are only one of the forms of solar activity. Flares, prominences and the solar magnetic field have cyclic variations. Since the Sun rotates every 27 days, we might expect terrestrial phenomena related to solar centres of activity to show a tendency to repeat at 27-day intervals. Such phenomena should also have an 11-year periodicity (Fig. 95), but a 22-year period in terrestrial phenomena could be related to the Sun *only* through the Sun's 22-year magnetic cycle. There *are* well-known events such as geomagnetic storms and short-wave radio fade-outs which show the 27-day period but these are associated only with the almost (but not completely) negligible parts of the Earth's atmosphere above about 80 km where the pressure is only 1/100 000 that at ground level. The well-known effects are certainly not associated with the lower atmosphere where our weather takes place.

What do we mean by 'weather'? The things of most general interest are rainfall and temperature, although meteorologists are concerned with many related properties, including pressure. Because of the great variability, extensive information is needed. This means records over as long a time and for as many stations as possible. Here is the first difficulty. Reliable sunspot records cover only from about 1700 AD onwards, and recordings of rain and temperature have been made for much shorter times in many parts of the globe. *Average data from over the Earth show little or no detectable variation which could be possibly related to solar activity.*

Around 1970, Russian astronomer E. R. Mustel drew attention to a curious phenomenon (latest results by Mustel and his collaborators V. E. Chertoprud and N. B. Moulukova are shown in Fig. 96). About

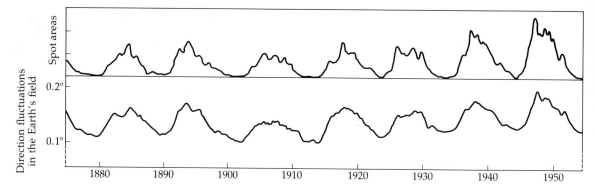

Fig. 95. *Sunspot areas and corresponding fluctuations in the direction of the Earth's magnetic field from 1874 to 1954. Although the changes in field direction are only a small fraction of a degree, they are obviously produced by some form of solar activity. We now know that this activity is the streams of particles associated with flares and large, stable coronal holes.*

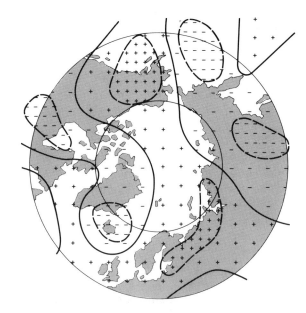

Fig. 96. *Russian astronomers E. R. Mustel, V. E. Chertoprud and N. B. Moulukova have found that, particularly in winter, the atmospheric pressure at ground level changes systematically some 3 days after the commencement of an isolated geomagnetic storm. Shown here are changes associated with sporadic storms produced by solar flares, in which the pressure rises over large parts of the northern hemisphere (+). The pressure falls in other large regions (−). Contours show where the changes reach ±0.5%; none reaches ±1%. Similar effects are produced by recurrent magnetic storms, associated with the 27-day rotation of large, stable coronal holes around the Sun.*

Fig. 97. *Mean variations in precipitation (rain and snow) in various latitude zones in the northern hemisphere, from 1885 to 1969. (The 50° to 60° zone refers to America only.) The broken line shows the corresponding variation in solar activity.*

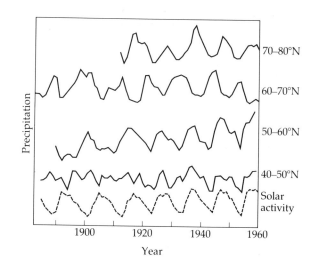

three days after the onset of a sporadic geomagnetic storm triggered by solar flares, or a recurrent storm associated with the 27-day rotation of large, stable coronal holes, systematic increases in atmospheric pressure of the order of 0.5% are found particularly in winter over large but still limited regions in the northern hemisphere, particularly above about latitude 55°. Other regions show systematic decreases in pressure. No such study has been possible for the southern

108

Fig. 98. Variation in rainfall from average (mm), when short-term fluctuations and very long-term trends are eliminated. Above, in Cairns, Queensland (17°S); below, Hobart, Tasmania (43°S). The rainfall variations are of exactly opposite phases, the minima at Cairns and maxima at Hobart occurring close to sunspot maximum.

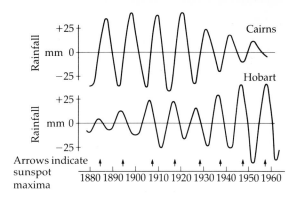

hemisphere, where data for the higher latitudes have been scanty until recently.

Before long, other interesting results started to appear. Fig. 97, giving results by John Xanthakis of Greece, shows the average fall of rain and snow over large zones in the northern hemisphere. There is a remarkable periodicity in the rainfalls over these large zones, apparently identical with the sunspot period. However, the precipitation and sunspot cycles are exactly in phase in the 70–80°N zone and 180° out of phase in the 60–70° zone. If the two zones are averaged, the 11-year periodicity would disappear. The 50–60°N zone behaved like the 60–70°N zone from 1890 to 1910, and then switched to be like the 70–80° zone.

Another interesting comparison between rainfall at different latitudes is due to Edward ('Taffy') Bowen in Australia (Fig. 98). Using smoothed rainfall data, he found from 80 years' records that at sunspot maximum there is a minimum in rainfall at Cairns (17°S) and a maximum at Hobart (43°S).

These results show that weather in large parts of the globe responds to the sunspot cycle, but that the sense of the response varies from one large area to another. Obviously, solar activity is not the main factor in producing rain, although it seems to have a systematic effect at the 10% level.

Much criticism has been aimed at the use of limited amounts of data in studying a possible relation between solar activity and weather. One possible way to obtain longer runs of weather data is to use tree rings which are laid down annually, thicker in wet years, thinner in drought years. However, much depends on the location and type of the tree, the nature of the ground, and its

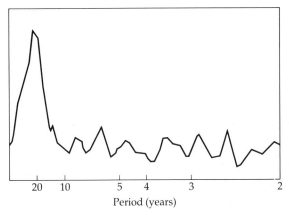

Fig. 99. Strengths of periods present in western-USA bristlecone pine tree rings. The strongest is 22 years. Other periods are of no statistical significance.

water retention. Averages are needed over large numbers of trees. The best-known centre for these studies is the Laboratory for Tree-Ring Research at the University of Arizona. Here they have concentrated on the bristlecone pine, the longest-living of all trees, using data covering the USA west of the Mississippi.

Neither in individual tree rings nor in the average data is a solar period *obvious*, but when properly analysed, the rings show a periodicity at 22 years, the solar magnetic period. The probability of this being due to chance is under 1%. No other period shorter than, say, 50 years is of any statistical significance. The rings are thinnest, and so droughts most likely, about 2 to 3 years after every second sunspot minimum, i.e. close to every second sunspot maximum.

109

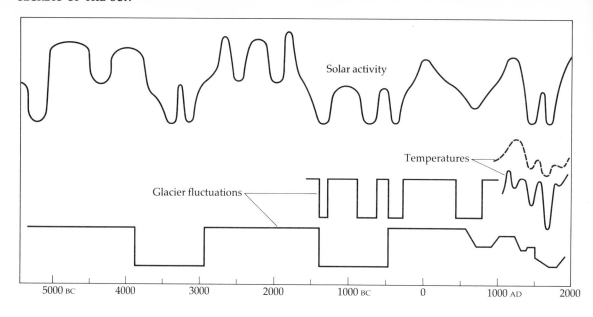

5000 BC	4000	3000	2000	1000 BC		0	1000 AD	2000

Fig. 100. Eddy's estimate of solar activity over the past 7500 years (top curve) compared with various estimates of climate. The two 'curves' of glacier fluctuations show epochs of warmer and cooler weather (downward is cooler). The broken curve shows the estimated mean annual temperature in England; the total range is about 1 °C. The continuous curve below it gives an index of the severity of winters in the Paris–London area. In general, strong solar activity corresponds to warmer temperatures and low activity to cooler temperatures.

Extrapolation beyond these records indicates that there should have been extensive droughts in USA west of the Mississippi around 1978 – and there were, actually in 1977!

Making use of tree rings as rainfall indicators, we can compare rainfall and the solar cycle back to the beginning of sunspot records. There is a striking relation between average sunspot activity and average drought-index from 1700 AD onwards. Even during the period of almost zero solar activity known as the Maunder minimum (1650 to 1700) for which the data are poor, the drought index, initially high, dropped continuously to its lowest value ever.

There is the possibility of extending the record of weather versus solar activity backwards to about 5500 BC, since bristlecone timber has now been dated covering this range of years, but how do we know what

solar activity was like at those times?

There is an indirect method of finding this, through the ratio of the ^{14}C and ^{12}C isotopes in the timber. There are two stable carbon isotopes (atoms having exactly the same chemical properties but different atomic weights); most have atomic weight 12 (^{12}C), and about 1% 13 (^{13}C). A rare radioactive isotope, ^{14}C, is produced by the bombardment of ordinary nitrogen by cosmic rays. This decays with a half-life of 5730 years. It is generally believed that, after correction for other known sources of variation, the residual fluctuations in the rate of producing ^{14}C can be traced to solar activity. Measurements of the ratio of the concentrations of ^{14}C and ^{12}C can thus be used to estimate the solar activity at the date of the sample.

Jack Eddy of the US High Altitude Observatory, who has been a leader in studying the history of the solar cycle, has prepared a diagram (Fig. 100) showing inferred solar activity over the past 7500 years and the corresponding temperature variations deduced from historical and recent geological evidence. It is clear that the relation found previously from studies of tree rings back to 1600 AD can be traced further back in time, though with reduced precision, for many thousands of years.

Fig. 101. A varve sample (length 22 cm) from a Late Precambrian glacial lake in the Flinders Ranges, South Australia. The rock shows annual layers whose widths indicate the amount of precipitation (rain and snow). At intervals of about 11 years there is a dark clayey band. Alternate 11-year intervals often show systematic fluctuations between dark bands spaced widely and closely. Most varve layers are flat rather than curved as here. These varves show that the climate in South Australia was sensitive to the sunspot cycle about 680 million years ago. They also show that the 11-year and 22-year cycles have scarcely changed during this time.

The best evidence of the effect of solar activity on weather is of a completely different type. George Williams has shown that varves, or annual layers, laid down in a Late Precambrian glacial lake in South Australia 680 million years ago contain a remarkable record of solar activity at that time (Figs 101 and 102). The 11-year cycle is strong, but for lengthy sequences covering perhaps 100 years or more there were alternate thin and thick 11-year groupings. The record then was more of a 22-year period. Such behaviour is in close agreement with what is known of the variation in strength of sunspot cycles over the past 300 years; a clear alternation has been present for about the past 130 years. Williams' initial analysis, covering 1760 years of the Late Precambrian era, shows that the 11-year cycle had a mean length of 11.2 years as compared with the modern 11.0 years. This extraordinary agreement shows that, at least in South Australia, weather was very responsive then to solar activity, and that the length of the sunspot cycle has not changed significantly over the past 680 million years.

Williams also discovered longer cycles: one of about 290 years which he named the Elatina cycle after the local rock formation, and one near 145 years (a second harmonic of the Elatina cycle). There is also a weaker cycle of about 90 years, probably identical with the Gleissberg cycle of about 80 years appearing in the modern sunspot record.

We should not be too worried by any apparent disagreements between Eddy's solar activity diagram and the varve data. There have been many other studies of various types aiming to recover past solar activity. Despite the poor accuracy usually involved, most of these have revealed cycles of roughly 300 and 150 years, comparable in length with those preserved in the varves. Moreover, detailed study of the Elatina Formation has only just begun. Williams has now drilled through the varve sequence, hoping to extend to some 20 000 years the ancient record of solar activity. From this we could expect information about the uniformity, or variability, of the Elatina cycle.

Another similar set of varve cycles has now been discovered in South Australia. They too confirm an impressive record of solar activity dating back to an Ice Age some 100 million years earlier.

An even older varve record has been claimed to show the 22-year solar magnetic cycle. A. F. Trendall has reported that rocks from the Precambrian Hamersley Group in Western Australia show strong banding at intervals averaging 23.3 years. They are about 2.5 billion years old, more than half the age of the Earth, and suggest that the magnetic cycle was affecting the weather even then. Little if any evolutionary change in period seems to have occurred over this great time span. It is possible that the strong relation between solar activity and weather found in the various varves was due to the Sun's magnetic field being much stronger then than now, but there is no way of testing this.

With three examples of conspicuous varve cyclicity already known, there are excellent prospects for further discoveries of varves containing useful information on solar activity during the Earth's long history.

In summary, there is strong evidence that solar activity has had major effects on weather in the remote past, and that it has been influencing weather over the last 1000 to 7500 years, though probably to a much reduced extent. We can assume confidently that it is doing so today, even if the problems of analysing short series of data, usually less than 100 years, make the detailed results for any locality unreliable. It is the task of the solar-terrestrial physicist to unravel the cause of the Sun's influence on weather, and its different effects in various parts of the world.

And what of the future? Can we find any way of predicting solar activity fluctuations sufficiently well to help long-term weather prediction? First, solar activity is responsible for only part of the Earth's weather. Charles Stockton, of the Laboratory for Tree-Ring Research, has suggested that only 20% of western USA droughts are related to the solar cycle. Secondly, while the dates of a sunspot cycle can be predicted a few decades ahead, its strength is usually uncertain until the cycle has started. Perhaps an extension of the theory outlined in Chapter 7 may be developed to help here, but no attempt has been made to do this yet.

It is tempting to use the longer-period varve cycles for purposes of prediction. There is the difficulty that it is more than 290 years since the middle of the Maunder minimum (around 1675 AD), whilst there has been no subsequent recurrence. This may well be due to some fluctuation in the length of the Elatina cycle. According to the varve record, the strongest maximum is followed by a fairly rapid descent to a low minimum. The Sun has had in 1958 and 1980 the strongest and next strongest maxima in the

Fig. 102. *Detail of Fig. 101 on which the individual annual deposits can be counted. What length was the average solar cycle in this sample?*

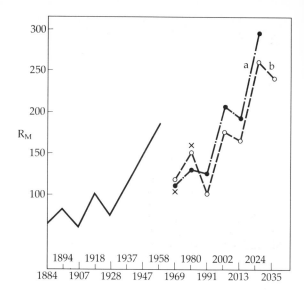

Fig. 103. The maximum sunspot number in successive cycles R$_M$, 1884 to 1958. The two broken curves (a and b) are Kopecký's predictions of future solar activity after 1958, using slightly differing assumptions. Crosses show the actual values of R$_M$ for the two cycles since 1958; they agree well with Kopecký's predictions.

past 370 years. Williams expects a rapid decline in the next few cycles.

There is, however, also a case for the opposite point of view. The most successful attempt to forecast strengths of future cycles has been made by the Czech astronomer M. Kopecký. His forecasts, which used data to 1958, are shown in Fig. 103. In the two subsequent cycles the Sun has behaved much as predicted. If it continued to do so until 2030 AD, solar activity would be at the highest level ever recorded.

Who will be right, Williams or Kopecký? We shall have to be patient for a few decades before the answer is known. The consequences are enormous. If solar activity increases greatly, there will be a hot period covering England and probably Europe early next century; western USA will have major droughts. Robert Vine's work in Australia suggests that this would be a time of severe bushfires, much more devastating than the record fires and droughts in the years near the strong 1980 sunspot maximum. Related work suggests the same thing for forest fires in Canada.

If solar activity decreases greatly, the opposite would apply. It would be cooler on the whole; but we do not know how temperatures would compare with those during the Maunder minimum, when the River Thames froze hard.

In either case, the changes are likely to have a major impact on crops and rainfall. As the world's economy is delicately poised and its population growing rapidly, countries with large populations marginally supported by food production, e.g. India, China and to a lesser extent even the USSR, need to take active and urgent steps to investigate the probable future course of solar activity and its effects on their food-producing regions.

Acknowledgements

Many of the pictures in this book have been provided directly by astronomers or observatories. In some cases friends of the author have made available some of their unique photographs. In others, the author has processed, either at Kitt Peak National Observatory or CSIRO Division of Applied Physics, photographs originating elsewhere so as to show particular phenomena optimally. The author expresses his gratitude and sincere thanks to all concerned for their generosity and help. The remaining photographs have been obtained by the author or under his supervision either at Kitt Peak National Observatory or CSIRO Division of Applied Physics.

The photograph on the dust jacket and Plate II are courtesy of W. C. Livingston. *Figs 1, 2* and *54* are courtesy of Mt Wilson and Las Campanas Observatories.

Figs 4, 5, 6 and *7* are enlarged extracts from originals by P. Janssen (*Ann. Obs. Meudon*, **1**, 1896).

Figs 8, 10, 63, 86 and *88* are courtesy of Association of Universities for Research in Astronomy (AURA) Inc., Sacramento Peak Observatory.

Fig. 9 is courtesy of AURA Inc., Sacramento Peak Observatory. Copyright © 1972 by D. Reidel Publishing Company, Dordrecht, Netherlands.

Fig. 11 is courtesy of and copyright © by National Geographical Society – Palomar Sky Survey. Reproduced by kind permission of the California Institute of Technology.

Fig. 13 is by the author using facilities provided by the Kitt Peak National Observatory.

Fig. 14 and *Plate IV* are by H. R. Gillett and the author using facilities provided by CSIRO Division of Applied Physics. *Fig. 14* is based on a spectrum secured by A. K. Pierce at Kitt Peak National Observatory.

Fig. 18, and *Plates VI* and *VIII* are by W. C. Livingston, courtesy of and copyright © 1982 by AURA Inc., Kitt Peak National Observatory.

Figs 19 and *31* are by J. M. Beckers, courtesy of AURA Inc., Sacramento Peak Observatory.

Fig. 20 is by G. W. Simon, courtesy of AURA Inc., Sacramento Peak Observatory.

Figs 21 and *22* are by the author, courtesy and copyright © 1982 by AURA Inc., Kitt Peak National Observatory.

Fig. 27 is by E. Fossat, G. Grec and M. Pomerantz, copyright © 1981 by D. Reidel Publishing Company, Dordrecht, Netherlands.

Fig. 28 is courtesy of Observatories du Pic-du-Midi et de Toulouse.

Fig. 29 is by R. E. Loughhead and R. J. Bray, courtesy *Australian Journal of Physics*.

Fig. 30 is courtesy and copyright © 1982 by D. Reidel Publishing Company, Dordrecht, Netherlands.

Figs 33 and *34* are courtesy of and copyright © 1972 by D. Reidel Publishing Company, Dordrecht, Netherlands.

Fig. 35 is by L. Cram and C. Zwaan, courtesy of AURA Inc., Sacramento Peak Observatory.

Figs 36 and *59* are courtesy of and copyright © 1974 by R. J. Bray and R. E. Loughhead.

Fig. 37 is by J. H. Piddington, courtesy of and copyright © 1976 by the IAU.

Figs 41, 42 and *44* are courtesy of and copyright © 1982 by AURA Inc., Kitt Peak National Observatory.

Fig. 45 and *Plate XIV* were processed by the author from KPNO daily magnetograms. Courtesy of and copyright © 1982 by AURA Inc., Kitt Peak National Observatory.

Fig. 46 is reproduced with permission from data supplied by the Science and Engineering Research Council, Royal Greenwich Observatory.

Fig. 47 is by J. Eddy, courtesy of and copyright © 1977 by D. Reidel Publishing Company, Dordrecht, Netherlands.

Figs 53, 83 and *84* are by R. B. Dunn, courtesy of AURA Inc., Sacramento Peak Observatory.

Figs 55 and *56* are courtesy of and copyright © 1974 by D. Reidel Publishing Company, Dordrecht, Netherlands.

Fig. 57 is by R. B. Dunn and J. Zirker, courtesy and copyright © 1973 by D. Reidel Publishing Company, Dordrecht, Netherlands.

Fig. 58 is courtesy of Lockheed Solar Observatory.

Figs 62 and *93* are courtesy of Big Bear Solar Observatory.

Fig. 64 is by V. Gaizauskas, courtesy of Ottawa River Solar Observatory, Herzburg Institute of Astrophysics, National Research Council of Canada.

Fig. 65 is by R. B. Dunn, courtesy of AURA Inc., Sacramento Peak Observatory. Copyright © 1974 by the IAU.

Fig. 68 is courtesy of Harvard College Observatory.

Figs 69 and *70* are courtesy of and copyright © 1978 by D. Reidel Publishing Company, Dordrecht, Netherlands.

Fig. 73 is by John L. Streete and Leon B. Lacey. Courtesy of High Altitude Observatory and Southwestern at Memphis, Tennessee.

Fig. 74 is by J. H. Rush, J. W. Firor, C. G. Lilliequist, L. B. Lacey and H. K. Hull. Courtesy of High Altitude Observatory.

Fig. 75 is courtesy of Los Alamos National Laboratory.

Fig. 76a is by K. Sheridan. Courtesy of CSIRO Division of Radiophysics and Astronomical Society of Australia.

Fig. 76b is an original photo obtained in the programme of the Solar Physics Group, American Science and Engineering, Inc., Cambridge, Massachusetts.

Fig. 77 is courtesy of the Naval Research Laboratory, Washington DC. Copyright © 1976 by D. Reidel Publishing Company, Dordrecht, Netherlands.

Fig. 78 is by K. L. Harvey, J. W. Harvey and S. F. Martin. Copyright © 1975 by D. Reidel Publishing Company, Dordrecht, Netherlands.

Fig. 79 is courtesy of G. W. Pneuman and R. Kopp. Copyright © 1971 by D. Reidel Publishing Company, Dordrecht, Netherlands.

Figs 81 and *85* are courtesy of the Institute for Astronomy, University of Hawaii.

Fig. 82 is courtesy of World Data Center A for Solar Terrestrial Physics.

Fig. 87 is courtesy of the Astronomical Society of Australia.

Fig. 89 is courtesy of S. R. Kane, copyright © 1974 by the IAU.

Fig. 91a is by M. K. McCabe and R. R. Fisher, courtesy of the Institute for Astronomy, University of Hawaii. Copyright © 1970 by D. Reidel Publishing Company, Dordrecht, Netherlands.

Figs 90, 91b and *92* are courtesy of CSIRO Division of Radiophysics and the Astronomical Society of Australia.

Fig. 94 is by R. Kopp and G. W. Pneuman, copyright © 1976 by D. Reidel Publishing Company, Dordrecht, Netherlands.

Fig. 95 is adapted from 'Sunspot and geomagnetic storm data 1874–1954', Royal Greenwich Observatory.

Fig. 96 is adapted from a diagram supplied by courtesy of E. R. Mustel, V. E. Chertoprud and N. B. Moulukova.

Fig. 97 is adapted from J. Xanthakis, *Proceedings of the First European Astronomy Meeting*, courtesy of Springer-Verlag, Heidelberg, West Germany.

Fig. 98 is courtesy of E. G. Bowen.

Fig. 99 is from a diagram by J. M. Mitchell, G. W. Stockton and D. M. Meko. Copyright © 1979 by D. Reidel Publishing Company, Dordrecht, Netherlands.

Fig. 100 is by J. Eddy (with minor modifications). Copyright © 1977 by D. Reidel Publishing Company, Dordrecht, Netherlands.

Fig. 102 is courtesy of R. N. Bracewell and J. Masterson, and of CSIRO Division of Radiophysics.

Fig. 103 is after M. Kopecký, courtesy of the Astronomical Institute of Czechoslovakia.

Plate I is courtesy of Carol Clarke.

Plate VII is by J. W. Harvey and T. L. Duvall. Courtesy of and copyright © 1982 by AURA Inc., Kitt Peak National Observatory.

Plates IX and *X* are by the author, based on magnetograms obtained by Kitt Peak National Observatory. Courtesy of and copyright © 1982 by AURA Inc., Kitt Peak National Observatory.

Plate XI is by the Victorian Tapestry Workshops. Courtesy of artist John Coburn and of The Speaker, Queensland Parliament.

Plate XII is courtesy of Virginia Sonett.

Plates XVI and *XV* were processed from photos supplied by courtesy of the Naval Research Laboratory, Washington DC. Copyright © 1982 by AURA Inc., Kitt Peak National Observatory.

Plate XVIII is courtesy of CSIRO Division of Radiophysics. Copyright © 1969 by D. Reidel Publishing Company, Dordrecht, Netherlands.

Plate XIX is courtesy of CSIRO Division of Radiophysics.